D0912686

Contrast analysis

Contrast analysis:

Focused comparisons in the analysis of variance

Robert Rosenthal
Harvard University

and

Ralph L. Rosnow
Temple University

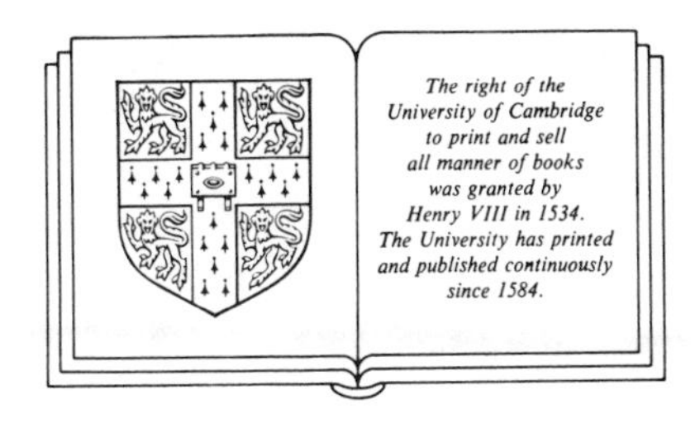

CAMBRIDGE UNIVERSITY PRESS

Cambridge
London New York New Rochelle
Melbourne Sydney

Published by the Press Syndicate of the University of Cambridge
The Pitt Building, Trumpington Street, Cambridge CB2 1RP
32 East 57th Street, New York, NY 10022, USA
10 Stamford Road, Oakleigh, Melbourne 3166, Australia

First published 1985

Printed in the United States of America

Library of Congress Cataloging in Publication Data

Rosenthal, Robert, 1933–

Contrast analysis.

Includes index.

1. Analysis of variance. 2. Psychometrics.
3. Social sciences – Statistical methods. I. Rosnow, Ralph L. II. Title.
BF39.R55 1985 519.5′352 84-23306
ISBN 0-521-30252-8 hard covers
ISBN 0-521-31798-7 paperback

To Julius and Hermine Rosenthal
and
Irvin and Rebecca Rosnow
For their Contrasts, their Analyses, and their Focus

Contents

Preface

This volume evolved from material developed to supplement our discussion of data analysis in *Essentials of Behavioral Research* (Rosenthal and Rosnow, 1984). Readers familiar with that text will recognize that the format of Chapter 2 of this volume closely corresponds (except for the addition of new examples) to the format of our discussion of contrast analysis (Chapter 23) in *Essentials*. Our purpose in fostering this similarity was to make it easier to use this volume in research methods courses as a collateral text or follow-up text to *Essentials*. We also envision its use for students of psychology, sociology, communication, education, and business as a collateral text to such standard texts in statistics as those by Hays; Keppel; Kirk; Myers; Welkowitz, Ewen, and Cohen; Winer. In research courses where both *Essentials* and this volume are used together, we recommend that students skip Chapter 23 in *Essentials* and instead read this volume in its entirety.

As in *Essentials*, our approach to the teaching of data analysis is intuitive, concrete, and arithmetic rather than rigorously mathematical. The statistical examples we employ are in all cases hypothetical, constructed specifically to illustrate the logical bases of the computational procedures. The numbers are neater than real-life examples usually tend to be, and there are fewer numbers in any single example than we would find in an actual data set. All of this material has been extensively pretested in our own classes and by several colleagues who used these or earlier versions of these chapters in their advanced undergraduate or graduate courses.

We are very much indebted to the many colleagues who provided comments on some or all of an earlier draft; and our thanks go to Pierce Barker, Bella DePaulo, Robin DiMatteo, Susan Fiske, Howie Friedman, Judy Hall, Dan Isenberg, Chick Judd, Dave Kenny, Charles Thomas, and Ed Tufte. The first author also thanks William G. Cochran, Jacob Cohen, Paul W. Holland, Frederick Mosteller, and Donald B. Rubin, who were influential in developing his philosophy of research and data analysis, and the National Science Foundation for its support of much of the research leading to the methodological developments described in this book. The second author is grateful for the support that has been forthcoming from Temple University in the form of the Thaddeus Bolton Professorship. Both authors express their appreciation to Blair Boudreau for her superb typing. We also thank David Serbun of McGraw-Hill and Susan Milmoe of Cambridge University Press for encouraging us to publish this material, and the following for generously granting permission to adapt the statistical tables that appear in the Appendix of this volume: the American Statistical Association, Houghton Mifflin Co., and McGraw-Hill.

This is our sabbatical volume - which is not to say that it was written during either of our sabbaticals, only that it is our seventh book together. In all seven instances we learned a great deal from each other and enjoyed ourselves greatly. We have also learned that the U.S. Postal Service and the telephone companies can nicely overcome physical distances. This is the third of our books for which MaryLu Rosenthal prepared the index and the seventh of our books for which she and Mimi Rosnow prepared and sustained the authors.

Robert Rosenthal
Ralph L. Rosnow

Postscript to preface. Long after our book was completed we learned of a brilliant unpublished paper by Robert Abelson that anticipated in spirit much of what we say in this volume. This discovery created ambivalence. On the one hand, it was discouraging to have had our own views so beautifully and so elegantly anticipated and by nearly a quarter of a century! On the other hand, it was very encouraging indeed to have found a paper that was so sympathetic to our approach: After all, it *was* by Robert Abelson! Readers wanting a sample of this early paper will find a brief sample to form the heart of our final chapter.

1

Why contrasts?

The nature of contrasts

Contrasts are significance tests of focused questions in which specific predictions can be evaluated by comparing these predictions to the obtained data. By a focused test (as opposed to an omnibus test), we mean any statistical test that addresses precise questions, as in any 1 *df* F test or in any t test. Omnibus tests, on the other hand, are tests of significance that address diffuse (or unfocused) questions, as in F with numerator $df > 1$ or in χ^2 with $df > 1$.

To illustrate, suppose a developmental psychologist interested in psychomotor skills had children at five age levels (11, 12, 13, 14, and 15) play a new outer space shoot-'em-up video game called Spear-Man. The object of the game is to fly around advancing ranks of demon spaceships from the planet Rho and to try to "spear" them with space bullets while trying to avoid space garbage, meteorites, and rockets that are flying toward you from several different directions. Table 1.1 shows the mean performance of 10 children at each of these five age levels, and Table 1.2 shows the overall analysis of variance computed on these data.

Table 1.1. *Mean performance scores at five age levels*[a]

		Age levels		
11	12	13	14	15
25	30	40	50	55

[a] $n = 10$ at each age level.

Table 1.2. *Analysis of variance of performance scores*

Source	*SS*	*df*	*MS*	*F*	*p*
Age levels	6500	4	1625	1.03	.40
Within	70,875	45	1575		

We see in Table 1.2 that F for age levels is far from significant ($p = .40$). Should we conclude that age was not an effective variable in this investigation? If we did so, we would be making a mistake, for Figure 1.1 (which plots the means of Table 1.1) clearly shows a linear relationship between age and performance. When we compute the correlation between these variables, we indeed find $r(3) = .992$, $p < .001$ (two-tailed)!

Why were we led astray by the results of the analysis of variance telling us that age did not make a difference, when we see clear and obvious results once we plot the means and compute r? The answer is that our omnibus F test addressed the question of whether there were *any* differences among the five groups, disregarding entirely the arrangement of the ages that constitute

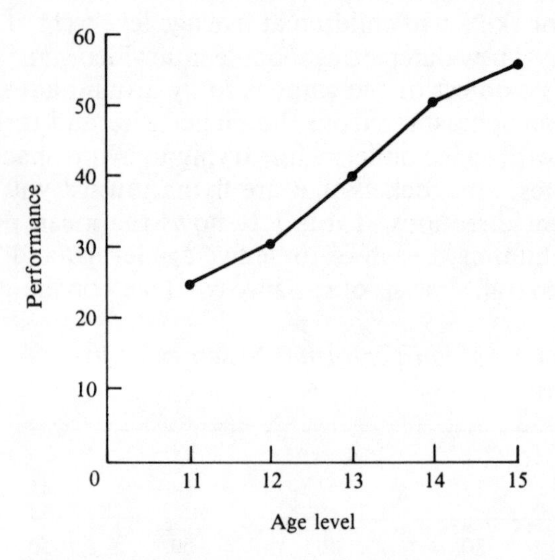

Figure 1.1. Means of Table 1.1 showing a linear relationship between age and performance.

the levels of the independent variable. We could rearrange the ages any way at all – 15, 14, 13, 12, 11 or 13, 12, 14, 15, 11 – and it would give us the same F as arranging them in the order previously shown. Our omnibus F test addresses a question actually of little interest to our researcher – the question is diffuse and unfocused. Our developmental researcher probably wanted to answer the question of whether psychomotor skills increased with age (i.e., showed a linear relationship) or first rose and then fell (i.e., showed a quadratic relationship), and so on. Our correlation addressed the specific question of whether performance increased linearly with age, and the answer was yes.

Comparing treatment means

To reiterate, contrast analysis permits us to ask focused questions of our data. Contrasts are comparisons employing two or more groups which we set up in such a way as to ensure that the results are compared to the predictions we make based on theory, hypothesis, or hunch. These predictions are expressed as lambda (λ) weights. The weights can take on any convenient numerical values as long as the sum of the weights ($\Sigma\lambda$) is zero for any given contrast. (We return to this idea in Chapter 2 and show how it applies in the example just discussed as well as in other illustrations of one-way analysis of variance.) What contrasts allow us to do is (a) to develop, at the outset, a number of focused questions we want to answer separately when analyzing our data (using *planned comparisons* instead of the overall analysis of variance and the omnibus F test) and (b) to use *incidental* or *post-hoc comparisons* to "snoop around" in our data after the overall analysis until we feel that we understand what the data actually show (Hays, 1963).

Previously, we noted as an example of a focused test any F with numerator $df = 1$ or any t test (and in this case, of course, $F = t^2$). Then we are comparing (or contrasting) only two means, $\overline{X}_1$ and $\overline{X}_2$. To be sure, we can also plan comparisons between means when there are more than two treatments (F with numerator $df > 1$), for example, three means:

$$\overline{X}_1 - \overline{X}_2 \qquad \text{and} \qquad \overline{X}_1 - \overline{X}_3 \qquad \text{and} \qquad \overline{X}_2 - \overline{X}_3$$

Suppose $\overline{X}_1$ is an experimental treatment group, $\overline{X}_2$ is a no-treatment control group, and $\overline{X}_3$ is a quasi-control group in which the subjects pretend (or simulate) being in the experimental group. We might well want to compare the average responses of each group with every other group. Alternatively, we might (after seeing there was no difference between the two control groups) want to combine the scores in $\overline{X}_2$ and $\overline{X}_3$ and compare the average responses in these two groups with those in $\overline{X}_1$; or else we might want to make up some other combination and average two of the three means and compare this average with the third mean, that is

$$\left(\frac{\overline{X}_1 + \overline{X}_2}{2}\right) - \overline{X}_3$$

and $$\left(\frac{\overline{X}_1 + \overline{X}_3}{2}\right) - \overline{X}_2$$

and $$\left(\frac{\overline{X}_2 + \overline{X}_3}{2}\right) - \overline{X}_1$$

In sum, the nature of the particular contrasts we choose to make - whether they be a series of comparisons between two treatment means or the average of two means compared with a third mean (as shown above) or a test for some trend in a given set of data - will depend on (a) our theory, hypothesis, or hunch and on (b) the type of data or research design we are using. In later chapters we shall examine a wide range of situations requiring different contrasts in the analysis of variance. Once we have mastered the use of contrasts to ask focused questions of our data (or of others' data, as discussed in Chapter 7), we will find that there are relatively few circumstances under which we will want to use omnibus F tests. What we will be getting in return for the small amount of computation required to employ contrasts is (a) very much greater statistical power and (b) very much greater clarity of substantive interpretation of research results.

Increased clarity of interpretation

It goes almost without saying that clarity of interpretation is essential in research, that is, to the extent that clarity is possible once we have subjected our data to proper examination (cf. Rosnow,

Table 1.3. *Mean scores of conservatives and liberals*[a]

Advertisement	Conservatives	Liberals	Means
Variation A	2.0	5.0	3.5
Variation B	2.0	1.0	1.5
Means	2.0	3.0	2.5

[a]n = 16 per cell.

1981). We will be talking more about interactions in later discussions of contrasts and a common example of the failure to subject data to proper examination occurs when researchers interpret interaction effects in the analysis of variance. Suppose we are comparing two variations of a political advertisement on a sample of 32 conservatives and 32 liberals. Table 1.3 shows the (posttreatment) opinion scores of the research participants, and Table 1.4 shows the analysis of variance of these data. In the published report, we could accurately state that there was a significant effect of the type of advertisement such that participants exposed to variation A expressed more positive opinions than participants exposed to B. We could also accurately state that there was no significant effect of the liberal versus conservative viewpoint of the participants on their expressed opinions regarding the advertisement. Finally, we might state – *but it would be wrong!* – that the significant interaction effect shown in Figure 1.2 (displaying the means of Table 1.3) demonstrates that liberals were more strongly influenced by advertisement A than by B while conservatives were unaffected by the type of advertisement.

Table 1.4. *Analysis of variance of opinion scores*

Source	*SS*	*df*	*MS*	*F*	*p*
Advertisements	64	1	64	4.0	.05
Respondents	16	1	16	1.0	—
Interaction	64	1	64	4.0	.05
Error term	960	60	16		

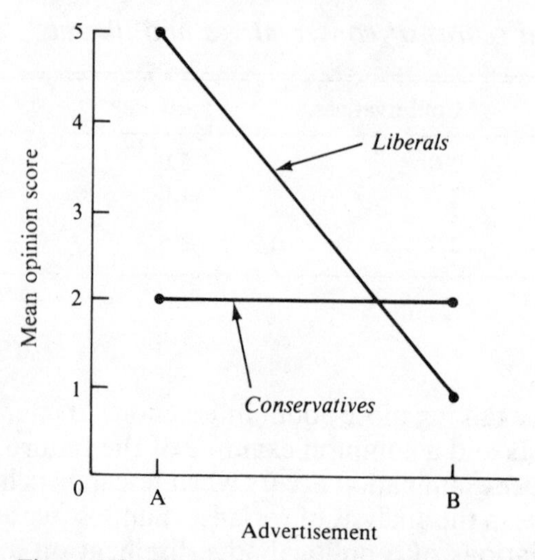

Figure 1.2. Overall results based on means of Table 1.3.

How have we erred? Figure 1.2 is a perfectly accurate display of the *overall* results of the study, in that it includes both main effects as well as the interaction. But it is not an accurate display of the interaction we believed was being depicted. To see what the interaction actually looks like, we would need to examine the data more closely. If we are to clarify the interaction accurately, then it is necessary for us to identify the residuals defining it. In this illustration all that is required is that we focus our data analysis by subtracting the grand mean along with the row and column effects from each condition of the experiment. Unfortunately, many researchers fail to do this, and instead base their interpretation (or rather misinterpretation) of the interaction on the overall means (i.e., the condition means) alone.

Table 1.5. *Row and column effects for means*

Advertisement	Conservatives	Liberals	Means	Row effects
Variation A	2.0	5.0	3.5	1.0
Variation B	2.0	1.0	1.5	−1.0
Means	2.0	3.0	2.5	
Column effects	−0.5	0.5		

Table 1.6. *Means "corrected for" row effects*

Advertisement	Conservatives	Liberals	Means	Row effects
Variation A	1.0	4.0	2.5	0
Variation B	3.0	2.0	2.5	0
Means	2.0	3.0	2.5	
Column effects	−0.5	0.5		

Since in Chapter 3 we return to the interpretation of interaction effects, it may be helpful if we review these calculations. Table 1.5 defines the particular row and column effects. Row effects are indicated for each row as the mean of that row minus the grand mean, and column effects are defined as the mean of each column minus the grand mean. Turning to Table 1.6, we see the original means "corrected for" row effects (that is, with the row effects removed), and in Table 1.7 we see these "corrected" means with the column effects now removed. In other words, we have decomposed the condition means shown in Table 1.3 to reveal the interaction effects to be the residual effects, or effects remaining after the lower-order effects of the rows and columns have been removed. The following summary table reviews what we have done so far:

	Interaction effect	=	Condition mean	−	Row effect	−	Column effect
Var A/Con	1.5	=	(2.0)	−	(1.0)	−	(−0.5)
Var B/Con	3.5	=	(2.0)	−	(−1.0)	−	(−0.5)
Var A/Lib	3.5	=	(5.0)	−	(1.0)	−	(0.5)
Var B/Lib	1.5	=	(1.0)	−	(−1.0)	−	(0.5)
Sums	10.0	=	10.0	−	0	−	0

Table 1.7. *Means "corrected for" row and column effects*

Advertisement	Conservatives	Liberals	Means	Row effects
Variation A	1.5	3.5	2.5	0
Variation B	3.5	1.5	2.5	0
Means	2.5	2.5	2.5	
Column effects	0	0		

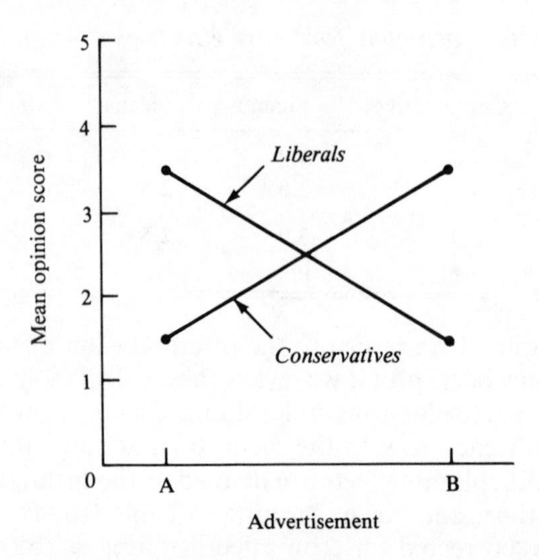

Figure 1.3. Interaction effect based on "corrected" means of Table 1.7.

Figure 1.3 displays the means shown in Table 1.7 (also in the column labeled "Interaction effect" above), that (with row and column effects removed) reveal the actual interaction effects. In this figure we see that the interaction actually shows that conservatives and liberals reacted in exactly opposite ways to the two types of propaganda. This is a very different conclusion than that based on the *overall* results depicted in Figure 1.2.

In most situations, we prefer to display the interaction effects freed of the grand mean as well as the row and column effects. To remove the grand mean from Table 1.7, we simply subtract 2.5 from every condition of the experiment. (For an extensive review of the computation and interpretation of interaction effects in the analysis of variance, see Chapter 21 in Rosenthal and Rosnow's *Essentials of Behavioral Research*, 1984.) Table 1.8 shows these results, which if plotted would be identical to those displayed in Figure 1.3 (except for changing the metric underlying the dependent variable. What we have done in subtracting the grand mean, row and column effects from the condition means can be summarized as follows:

	Interaction effect	=	Condition mean	−	Row effect	−	Column effect	−	Grand mean
Var A/Con	−1.0	=	(2.0)	−	(1.0)	−	(−0.5)	−	(2.5)
Var B/Con	1.0	=	(2.0)	−	(−1.0)	−	(−0.5)	−	(2.5)
Var A/Lib	1.0	=	(5.0)	−	(1.0)	−	(0.5)	−	(2.5)
Var B/Lib	−1.0	=	(1.0)	−	(−1.0)	−	(0.5)	−	(2.5)
Sums	0	=	10.0	−	0	−	0	−	10.0

Significance and effect size

In the example that began this chapter, we saw how an omnibus *F* test tells almost nothing except to emphasize overall significance of results. However, even when we do comparisons between treatment means, it is a good idea not to base our conclusions only on the statistical significance of our findings. Significance reveals only part of the overall picture. There will almost always be two kinds of information we want to have for each research question we address with planned or post-hoc comparisons: (a) the size of the effect and (b) its statistical significance. This theme can be expressed by a fundamental conceptual equation (Rosenthal and Rosnow, 1984):

$$\text{Significance test} = \text{size of effect} \times \text{size of study}$$

This equation tells us that, for any given size of effect (e.g., r, r^2, $r/\sqrt{1-r^2}$, or $r^2/(1-r^2)$) and size of study (e.g., N, df, $\sqrt{N}$, $\sqrt{df}$), there will be a corresponding test of significance.

Table 1.8. *Means "corrected for" row, column, and grand mean effects*

Advertisement	Conservatives	Liberals	Means	Row effects
Variation A	−1.0	1.0	0	0
Variation B	1.0	−1.0	0	0
Means	0	0	0	
Column effects	0	0		

Another way of saying this is that every test of significance (e.g., t or F) is made up of two components, the size of the effect and the size of the study. So, for example, when there are two means to be compared and the size of the effect of the independent variable is indexed by some correlation, the general relationship above can be rewritten as

$$t = \frac{r}{\sqrt{1 - r^2}} \times \sqrt{df}$$

or as

$$F = \frac{r^2}{1 - r^2} \times df$$

In the following chapters we shall refer both to estimates of the size of the relationship and its statistical significance as we explore various ways in which comparison procedures can be used in the analysis of variance. It is possible, of course, to do focused tests in the context of other analytic procedures, for example, within the framework of meta-analysis procedures that are used for comparing the significance levels or effect sizes of several research studies. When research studies are compared as to their significance or their effect sizes by focused tests, or contrasts, we learn whether the studies differ significantly among themselves in a theoretically predictable or meaningful way. Suppose we have a given set of p values for studies of teacher expectancy effects; we might want to know whether results from younger children show greater degrees of statistical significance than do results from older children (Rosenthal & Rubin, 1978a, 1978b). Meta-analytic procedures are extensively discussed elsewhere (Rosenthal, 1984), while in this volume we are concerned primarily with the essentials of comparison procedures in the analysis of variance.

2
One-way analysis

Basic computations

In Chapter 1, we began with an example of a one-way analysis of variance: the study by the developmental psychologist interested in psychomotor skills who had children at five age levels play a new video game. The means (see Table 1.1) were given as 25, 30, 40, 50, and 55; each mean based on an n of 10. Using these data we now see how contrasts are computed within the context of a one-way analysis of variance. Given the following formula, which shows the computation of a contrast in terms of a sum of squares for the single *df* test being made (Snedecor & Cochran, 1967, p. 308), we will show how contrasts can easily be computed within this context:

$$MS \text{ contrast} = SS \text{ contrast} = \frac{L^2}{n\Sigma\lambda^2}$$

where L = sum of all condition totals (T), each of which has been multiplied by the lambda weight (λ) called for by our theory, hypothesis, or hunch; that is,

$$L = \Sigma[T\lambda] = T_1\lambda_1 + T_2\lambda_2 + T_3\lambda_3 + \cdots + T_k\lambda_k$$

where k = number of conditions; n = number of observations in each condition, given equal n per condition (later we discuss the condition of unequal n's); and λ's = weights required by our theory, hypothesis, or hunch, such that $\Sigma\lambda = 0$. Since contrasts are based on only 1 *df*, we see in the first formula above that the sum of squares is indicated as identical to the mean square. For our F test, then, we simply divide the mean square by the appropriate error term.

Let us now apply this formula to the data of our example. To get the required values of T, we multiply our means by n, which gives us 250, 300, 400, 500, and 550. We now need a lambda weight for each T. Suppose our developmental psychologist's prediction was that there would be a linear trend, that is to say, a regular increment of performance on the video game for every regular increment of age. In this case, we might think of using age levels as our λ's, that is, 11, 12, 13, 14, 15. Unfortunately, the sum of these numbers is not zero as required, but 65. However, that is easy to rectify. We can simply subtract the mean age level of 13 from each of our λ's, which gives us: $(11 - 13)$, $(12 - 13)$, $(13 - 13)$, $(14 - 13)$, and $(15 - 13)$. We now have a set of weights that does sum to zero: -2, -1, 0, $+1$, $+2$. If you look at Table A.1 in the Appendix, you will see these same weights (for $k = 5$ conditions). You will also see additional weights given for quadratic and cubic orthogonal (i.e., independent) trends, curves, or polynomials (algebraic expressions of two or more terms). Using this table saves us the trouble of having to calculate these weights, and later in this chapter we shall describe these orthogonal polynomials in more detail.

What we have so far is shown in Table 2.1. What we still need are L^2 and $\Sigma\lambda^2$; and since $L = \Sigma[T\lambda] = 800$, then $L^2 = (800)^2$. We compute $\Sigma\lambda^2$ from Table 2.1 to be: $(-2)^2 + (-1)^2 + (0)^2 + (+1)^2 + (+2)^2 = 10$. So, given $n = 10$, we have

$$\frac{L^2}{n\Sigma\lambda^2} = \frac{(800)^2}{10(10)} = 6400 = SS \text{ contrast} = MS \text{ contrast}$$

Our F test for this contrast is computed by dividing it by the mean square for error of our analysis of variance previously shown in

Table 2.1. *Basic data for computing linear trend*

Age level	11	12	13	14	15	Σ
Mean	25	30	40	50	55	
T	250	300	400	500	550	2000
λ	-2	-1	0	$+1$	$+2$	0
$T\lambda$	-500	-300	0	500	1100	800

Table 1.2, and we find $F(1,45) = 6400/1575 = 4.06$. Table A.4 in the Appendix shows this F value to be significant at $p = .05$.

All F's used to test contrasts have only 1 *df* in the numerator, and therefore we can compute t simply by taking the square root of our F in case we want to make a one-tailed test. In the present case, we find $t(45) = 2.01$, which is significant at $p = .025$ one-tailed (see Table A.3 in Appendix).

A characteristic of contrast sums of squares is that they are identical whether we employ a given set of weights or their opposites (i.e., the weights multiplied by -1). Figure 2.1 shows the predictions of two hypothetical studies with exactly opposite contrast weights: (a) $+2, +1, 0, -1, -2$ and (b) $-2, -1, 0, +1, +2$. Using (a) instead of (b) in the preceding example would still give us the same results, namely, *SS* contrast $= 6400$ and $F(1,45) = 4.06$, $p = .05$. This p value, even though one-tailed in the distribution of F (in that it refers only to the right-hand portion of the F distribution), is two-tailed with respect to the hypothesis that performance increases with age. When we compute one-tailed t tests from the square root of F, we must be careful that the results do actually support our prediction and not its opposite. To

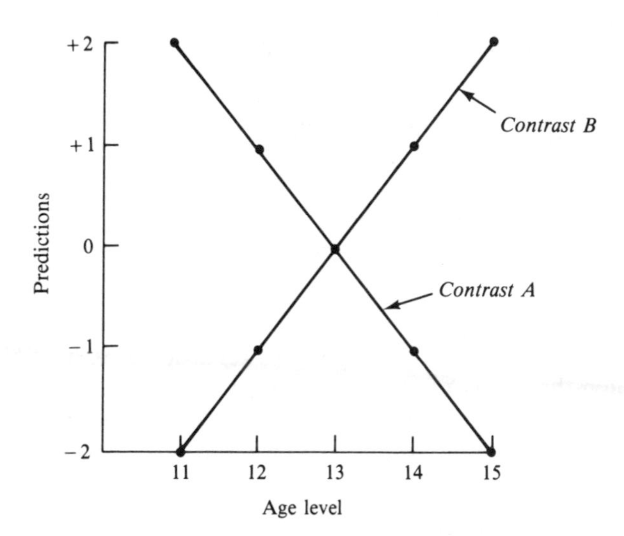

Figure 2.1. Display of predicted results with opposite contrast weights: (a) $+2, +1, 0, -1, -2$ and (b) $-2, -1, 0, +1, +2$.

minimize error, we give t a positive sign when the result is in the predicted direction (e.g., performance improves with age) and a negative sign when the result is in the opposite direction (e.g., performance worsens with age).

In our discussion at the end of Chapter 1, we noted that we almost always want to know the size of the effect as well as its statistical significance. In this case we can use r to estimate the size of the effect of the linear relationship between performance and age, where

$$r = \sqrt{\frac{(df \text{ numerator})F}{(df \text{ numerator})F + df \text{ denominator}}}$$

$$= \sqrt{\frac{t^2}{t^2 + df}}$$

$$= \sqrt{\frac{4.06}{4.06 + 45}} = .29$$

Here, then, the correlation between age level and average performance level ($r = .29$) is of moderate size. An alternative computational formula for the effect size is:

$$r = \sqrt{\frac{SS \text{ contrast}}{SS \text{ contrast} + SS \text{ error}}}$$

$$= \sqrt{\frac{6400}{6400 + 70{,}875}} = .29$$

where *SS* error was given in Table 1.2. As a terminological note we might add here that r is sometimes regarded as a special case of *eta* such that an *eta* is called r only when it estimates an effect size for a contrast so that $df = 1$. Whatever usage may be encountered, we want to emphasize that the relationship

$$\sqrt{\frac{(df \text{ numerator})F}{(df \text{ numerator})F + df \text{ denominator}}} = \sqrt{\frac{t^2}{t^2 + df}}$$

holds only when df numerator $= 1$.

You may recall from our discussion of these results in Chapter 1 that, when we computed the correlation between these performance means and the five age levels, we found $r(3) = .992$. That finding was, in fact, consistent with what we saw when we plotted the means as a function of age level in Figure 1.1. We obtain this same value if we now divide the *SS* contrast by the total *SS* between age groups and take the square root, that is,

$$\sqrt{\frac{6400}{6500}} = .992.$$

This r, you will remember, alerted us that we were making an error by ignoring the increasing nature of age. While $r = .992$ does a good job of estimating the correlation of age and performance for the mean age and mean performance of our groups of children ($n = 10$ per group), it is a poor estimate of the relationship between the individual child's age and performance. (In general, it is usually the case that correlations based on groups or other aggregated data are higher than those based on the original nonaggregated data.)

Further examples

Continuing with our example, suppose our developmental psychologist was confident only that 15-year-olds would be superior to 11-year-olds. The weights now chosen (λ's) might look as follows:

Age level	11	12	13	14	15	Σ
T	250	300	400	500	550	2000
λ	−1	0	0	0	+1	0
$T\lambda$	−250	0	0	0	+550	300

The *SS* contrast would now be:

$$\frac{L^2}{n\Sigma\lambda^2} = \frac{(300)^2}{10(2)} = 4500 \text{ (also} = MS \text{ contrast)}$$

This value, when divided by the mean square for error of our earlier analysis of variance (Table 1.2), yields: $F(1,45) = 4500/1575 = 2.86$, or $t(45) = 1.69$, $p = .05$, one-tailed.

Comparing the 15-year-olds to the 11-year-olds is, of course, something we could also do with a simple t test:

$$t = \frac{M_1 - M_2}{\sqrt{\left(\frac{1}{n_1} + \frac{1}{n_2}\right) MS \text{ error}}} = \frac{55.0 - 25.0}{\sqrt{\left(\frac{1}{10} + \frac{1}{10}\right) 1575}} = 1.69,$$

$df = 45$ (the df associated with the mean square error), $p = .05$, one-tailed. Comparing this ordinary t test with the contrast t test shows them to be identical, as they should be.

Now suppose that our developmental psychologist hypothesized that both the 11- and 12-year-olds would score significantly lower than the 15-year-olds. The researcher might test this idea by choosing weights (λ's) as follows:

Age level	11	12	13	14	15	Σ
T	250	300	400	500	550	2000
λ	−1	−1	0	0	+2	0
$T\lambda$	−250	−300	0	0	+1100	550

Keep in mind that our λ's must sum to zero, so that the $\lambda = +2$ of the 15-year-olds is needed to balance the −1 and −1 of the 11- and 12-year-olds.

The *SS* contrast would now be:

$$\frac{L^2}{n\Sigma\lambda^2} = \frac{(550)^2}{10(6)} = 5041.67 = MS \text{ contrast}$$

which when divided by the mean square for error of our earlier analysis of variance (Table 1.2), yields $F(1,45) = 5041.67/1575 = 3.20$, or $t(45) = 1.79$, $p = .04$, one-tailed.

Had we decided to compute a simple t test between the mean of the 11- and 12-year-olds [(25 + 30)/2 = 27.5, with $n = 20$] and the mean of the 15-year-olds (55, with $n = 10$), we would do as follows:

$$t = \frac{M_1 - M_2}{\sqrt{\left(\frac{1}{n_1} + \frac{1}{n_2}\right) MS \text{ error}}}$$

$$= \frac{55 - 27.5}{\sqrt{\left(\frac{1}{10} + \frac{1}{20}\right) 1575}}$$

$$= 1.79,$$

$$df = 45, \qquad p = .04, \text{ one-tailed}$$

Once again, we note that the two methods of computing t give us identical results as expected.

Unequal n per condition

So far in our discussion of contrasts we have assumed equal n per condition. But suppose we had a study consisting of five conditions in which the n's were 2, 8, 12, 14, and 14. One way to deal with the problem might be to discard a random subset of the units of each condition until the sample sizes of all conditions are equal. However, that would be terribly wasteful and would also reduce power substantially. A better procedure would be to employ an unweighted means approach, in which we redefine n as the harmonic mean and therefore also redefine T (Rosenthal & Rosnow, 1984).

Let us go back to our basic formula for computing SS contrast:

$$\frac{L^2}{n\Sigma\lambda^2}$$

This can be rewritten as

$$\frac{(\Sigma T\lambda)^2}{n\Sigma\lambda^2}$$

We now redefine n so it becomes the harmonic mean of the n's, and T so it becomes the mean of the condition multiplied by the harmonic mean of the *n*'s:

$$\text{redefined } n = \frac{k}{\sum \frac{1}{n}} = n_h$$

where n_h symbolizes the harmonic mean, k is the number of conditions, $\sum \frac{1}{n}$ is the sum of the reciprocals of the n's, and redefined $T = Mn_h$ where M is the mean of a condition and n_h is n as redefined above. In our study where the n's are 2, 8, 12, 14, 14, the arithmetic mean n would be 10 but the harmonic mean n would be 5.87:

$$n_h = \frac{k}{\sum \frac{1}{n}} = \frac{5}{\frac{1}{2} + \frac{1}{8} + \frac{1}{12} + \frac{1}{14} + \frac{1}{14}} = 5.87$$

Although the redefined n and redefined T are required when n's are unequal, in fact it is always appropriate to employ them in our calculations. When we have equal n's the arithmetic mean n and the harmonic mean n will be identical, because

$$n_h = \frac{k}{\sum \frac{1}{n}} = \frac{k}{k\left(\frac{1}{n}\right)} = n \quad \text{when } n\text{'s are equal}$$

Orthogonal contrasts

In a set of research results based on k conditions, we can compute up to $k - 1$ contrasts, each of which is orthogonal to (uncorrelated with) every other contrast. Contrasts are orthogonal when the correlation between them is zero, and the correlation will be zero when the sum of the products of the corresponding weights (λ's) is zero. For example, the following two sets of contrast weights, drawn from Table A.1 ($k = 6$) in the Appendix, are orthogonal:

Contrast	A	B	C	D	E	F	Σ
λ_1 set	−5	−3	−1	+1	+3	+5	0
λ_2 set	+5	−1	−4	−4	−1	+5	0
Product of $\lambda_1 \times \lambda_2$	−25	+3	+4	−4	−3	+25	0

These sets of contrast weights represent six points on a straight line (λ_1 set) and six points on a U-shaped function (λ_2 set). The third row (product of $\lambda_1\lambda_2$) shows the result of multiplying one set with another. We conclude that contrast sets 1 and 2 are orthogonal to each other since we see that the sum of the products (row 3) is equal to zero.

In parts (a) and (b) of Figure 2.2 we see the λ_1 set and the λ_2 set plotted to show what the linear and quadratic trends look like, and in part (c) we see what a cubic trend looks like. The weights corresponding to all the data points are listed in Appendix Table A.1 for $k = 6$ conditions. In this figure we observe that the linear trend shows a consistent gain, but it could also show a consistent loss. We observe that the quadratic trend shows a change in direction from down to up in a U curve, but it could also show a change from up to down in a ∩ curve (an upside-down U). Cubic trends, which can be assessed when there are four or more conditions, show two changes of direction from up to down to up (or down to up to down).

Referring back to the contrasts λ_1 and λ_2 in the previous table, we can plot the sum of these weights to show what a curve containing both linear and quadratic components looks like. Figure 2.3 shows these summed weights, that is, a curve with strong linear and strong quadratic components. In most real-life applications, we usually find combinations of linear and nonlinear results as opposed to the more idealized results shown in Figure 2.2.

Appendix Table A.1 thus provides us with a particularly useful set of orthogonal contrasts based on the coefficients of orthogonal polynomials (curves or trends). This table should be considered whenever the k conditions of the study can be arranged from the smallest to the largest levels of the independent variable, as is the case when age levels, dosage levels, learning trials, or other ordered levels comprise the independent variables.

We have noted that it is possible to compute up to $k - 1$ orthogonal contrasts among a set of k means or totals. Thus if we had four conditions, we could compute three orthogonal contrasts, each based on a different polynomial or trend, the linear, quadratic, and cubic. The sums of squares of these three contrasts would add up to the total sum of squares among the four conditions. However, although there are only $k - 1$ orthogonal contrasts in a given set such as those based on orthogonal polynomials, there

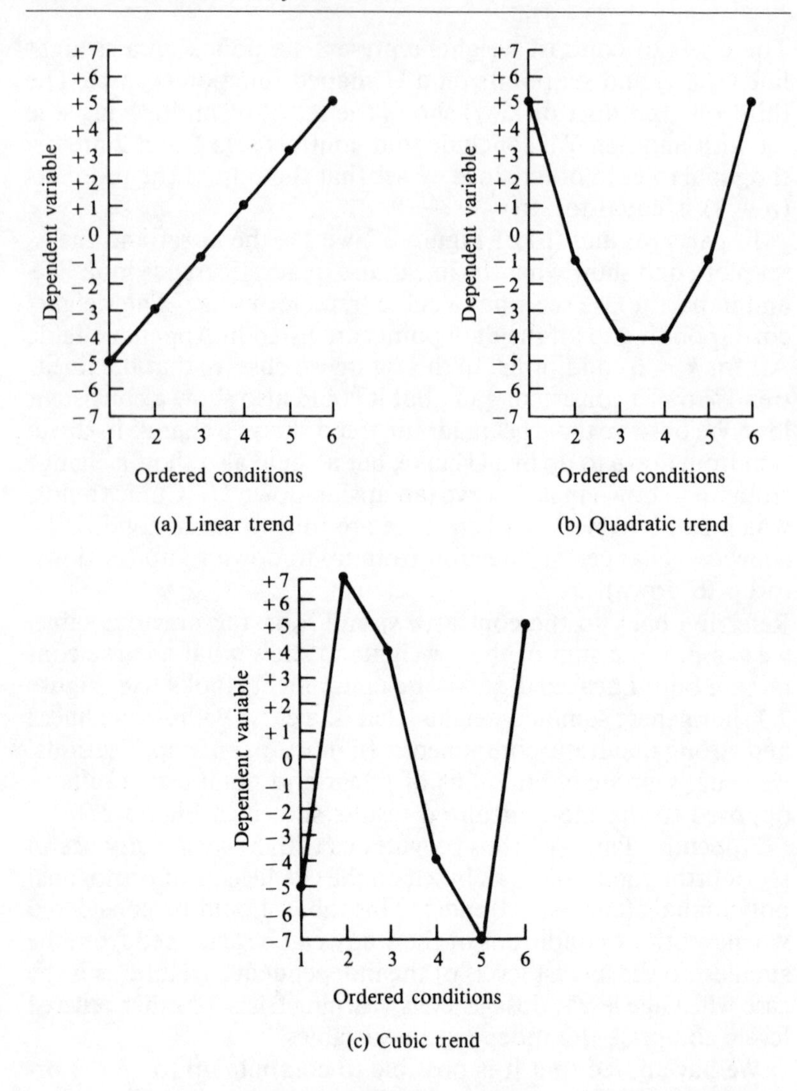

Figure 2.2. Results of three hypothetical studies that were (a) perfectly linear, (b) perfectly quadratic, and (c) perfectly cubic.

is an infinite number of *sets* of contrasts that could be computed, each of which is made up of $k - 1$ orthogonal contrasts. This idea of an infinite number of sets of contrasts is easy to grasp if we think of the endless number of contrast weights that could

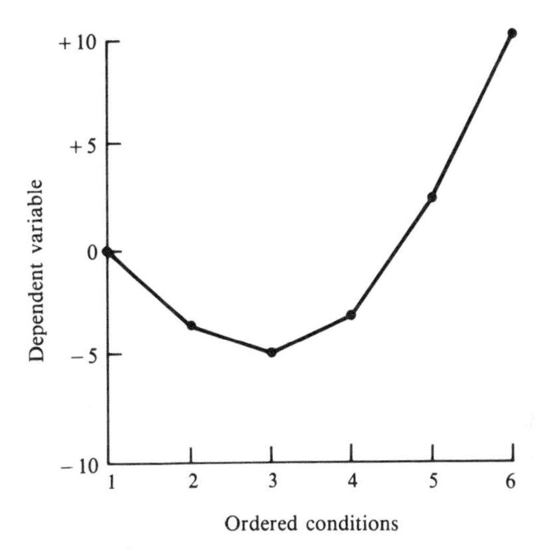

Figure 2.3. Curve showing both linear and quadratic components.

be assigned to as few as three means or totals, that is, any one λ could range from $-\infty$ to $+\infty$. The *sets* of contrasts, however, would not be orthogonal to one another. For example, in the following contrasts, Set I is comprised of mutually orthogonal contrasts, as is Set II, but none of the three contrasts in Set I is orthogonal to any of the contrasts of Set II:

	Contrast Set I					Contrast Set II			
	A	B	C	D		A	B	C	D
λ_1	−3	−1	+1	+3	λ_1	−1	−1	−1	+3
λ_2	+1	−1	−1	+1	λ_2	−1	−1	+2	0
λ_3	−1	+3	−3	+1	λ_3	−1	+1	0	0

Nonorthogonal contrasts

While there is some advantage to employing orthogonal contrasts – in that each contrast addresses a fresh and nonoverlapping question – there is no reason not to employ correlated (nonorthogonal) contrasts. One valuable use of nonorthogonal contrasts is in the comparison of certain plausible rival hypotheses. Suppose we tested children at grades 2, 5, 8, and 11. One plausible hypothesis (prediction 1) is for a constant rate of improvement with age, while a plausible rival hypothesis (prediction 2) predicts only that those in grade 11 will differ from all younger children. Table 2.2 shows the hypothetical results of this research, the contrast weights used to test each hypothesis, the sums of squares associated with each contrast, and the sums of squares between all conditions ($df = 3$).

Both contrasts do a good job of fitting the data, with SS_1 taking up $50/60 = 83$ percent of the between conditions SS, and SS_2 taking up $53.3/60 = 89$ percent of the between conditions SS. Prediction 2 did a little better than prediction 1, but not enough better to make us give up hypothesis 1. That both hypotheses did well should not surprise us too much, since the correlation between the weights representing the two hypotheses was quite substantial ($r = .77$). We might have had a third hypothesis that predicted that 2nd graders and 11th graders would differ most but that 5th and 8th graders would not differ from one another. That prediction would be expressed by λ's of $-1, 0, 0, +1$, and would have been correlated .95 with prediction 1 and .82 with predic-

Table 2.2. *Hypothetical results of research in which prediction 1 is a constant rate of improvement and prediction 2 is that one grade differs*

	Grade 2	Grade 5	Grade 8	Grade 11	Σ
Means	5.0	5.0	6.0	8.0	24.0
Prediction 1 λ's	-3	-1	$+1$	$+3$	0
Prediction 2 λ's	-1	-1	-1	$+3$	0
(with $n = 10$ at each age level)					

$$SS_1 = \frac{L^2}{n\Sigma\lambda^2} = \frac{(\Sigma T\lambda)^2}{n\Sigma\lambda^2} = \frac{[50(-3) + 50(-1) + 60(+1) + 80(+3)]^2}{10\ [(+3)^2 + (+1)^2 + (-1)^2 + (-3)^2]} = 50$$

$$SS_2 = \frac{L^2}{n\Sigma\lambda^2} = \frac{(\Sigma T\lambda)^2}{n\Sigma\lambda^2} = \frac{[50(-1) + 50(-1) + 60(-1) + 80(+3)]^2}{10\ [(-1)^2 + (-1)^2 + (-1)^2 + (+3)^2]} = 53.3$$

$$SS \text{ between conditions} = \frac{(50)^2}{10} + \frac{(50)^2}{10} + \frac{(60)^2}{10} + \frac{(80)^2}{10} - \frac{(240)^2}{40} = 60$$

tion 2. The *SS* contrast for this set of weights would be 45, accounting for 75 percent of the total variance among conditions.

Relationships: Linear and nonlinear

In this chapter, we have looked at examples of both linear and nonlinear relationships in the context of the one-way analysis of variance. It has been argued by various theorists that assumptions concerning the fundamental simplicity of the universe and consequent appeal to economy and simplicity in the explanatory efforts of science have led to a preference for linear laws and relationships (e.g., Battig, 1962; Luchins & Luchins, 1965). In fact, many researchers have necessarily limited themselves to the consideration of straight-line relationships by virtue of their having examined only two levels of the independent variable. There are, of course, important nonlinear relationships in the behavioral and social sciences. For instance, in the field of neuropsychology, it has been repeatedly found that the shape of the curve relating level of performance to level of activation is that of an inverted U (cf. Malmo, 1959; Schultz, 1965). In social psychology, researchers have become increasingly sensitive to the possibility of a large number of nonlinear effects of social influence variables (cf. Converse, 1982; Converse & Cooper, 1979; Eagly, 1981, p. 178; McGuire, 1968; Rosnow, 1980). In the area of experimental psychology, investigators have dealt with the analysis of curves that result when the difference between experimental treatments involves a scaled independent variable; for example, learning curves, extinction curves, dark adaptation curves, time-to-attain-a-level-of-dark-adaptation as a function of the luminance of the preadaptation field, response rate as a function of the amount of reinforcement, and other variables (Grant, 1956). Such an orientation is also compatible with recent epistemological developments in psychology which emphasize the idea of cycles and other complex relationships (Rosnow, 1978, 1981, 1983), though it is true that some earlier theorists were also interested in complex functional relationships even without the baggage of epistemology (e.g., Thorndike, 1918). In our discussion now, and in subsequent chapters, we delve into the possibility of both linear and nonlinear relationships in the context of more complex analyses of variance.

3

Two-factor studies

Contrasts for main effects

In this chapter we explore the decomposition of the main effects and interactions of two-way analyses of variance into appropriate and informative contrasts. We begin with the results of a hypothetical experiment on the treatment of depression (see Table 3.1). We see that there were four levels of the treatment factor and three levels of the age factor. There were 40 patients each in the older, middle-aged, and younger samples, and 10 of these patients were randomly assigned to each of the four treatment conditions. Two of the treatment groups were hospitalized and two were not. Within the former conditions, one group was treated

Table 3.1. *Hypothetical results of an experiment in treating depression (Data are sums of benefit scores; n = 10)*

	Nonhospitalization		Hospitalization		
Patient age	Psychotherapy A	Companionship program B	Traditional C	Milieu therapy D	Σ
Old	5 (10[a])	7	2	2	16 (40[a])
Middle age	9	7	4	8	28
Young	7	7	0	2	16
Σ	21 (30[a])	21	6	12	60 (120[a])

[a]These entries represent the number of observations upon which the total is based.

Table 3.2. *Traditional analysis of variance of data of Table 3.1*

Source	*SS*	*df*	*MS*	*F*	*p*
Treatment	5.40	3	1.80	36.00	very small
Age	2.40	2	1.20	24.00	very small
Interaction	1.60	6	.27	5.34	.001
Within	5.40	108	.05		

traditionally while the other received milieu (environmental) therapy. Within the latter conditions, one group was given psychotherapy while the other was given a new companionship program. In this table we see the *sums* of the 10 observations for each of the 12 combinations of treatment and age. In Table 3.2 we find the traditional two-way analysis of variance of these data, an analysis we regard as merely preliminary.

For our contrast analysis, we may want to start by computing a set of orthogonal contrasts among the four levels of treatments, though it is not required that our contrasts be orthogonal. There are three questions we wish to address by means of these contrasts: (a) Does hospitalization make a difference? (b) Given nonhospitalization, does psychotherapy differ from the companionship program? and (c) Given hospitalization, does traditional treatment differ from milieu therapy? Table 3.3 shows the treatment totals (taken from Table 3.1), the weights (λ) assigned for each of our three orthogonal contrasts, and the product of the totals (T) $\times$ the weights ($T\lambda$). Note that the weights are assigned according to the particular questions to be answered. For λ_1, the weights are consistent with question (a), which contrasts hospitalization (traditional and milieu) with nonhospitalization (psychotherapy and companionship). For λ_2, the weights are consistent with question (b), which contrasts only the two nonhospitalization treatments, therefore assigning zero weights to the irrelevant treatments. For λ_3, which addresses question (c), this is reversed, with the zero weights now assigned to the nonhospitalization treatments.

The right-hand column of Table 3.3 provides the sums of the preceding four column entries. The first value (60) is the grand

Table 3.3. *Treatment totals and contrast weights*

	Nonhospitalization		Hospitalization		
	Psychotherapy	Companionship	Traditional	Milieu	Σ
Totals (T)	21 [30a]	21 [30a]	6 [30a]	12 [30a]	60 [120a]
Nonhosp. vs. hosp.: λ_1	+1	+1	−1	−1	0
Psy. vs. comp.: λ_2	+1	−1	0	0	0
Trad. vs. mil.: λ_3	0	0	−1	+1	0
Totals × λ_1	21	21	−6	−12	24
Totals × λ_2	21	−21	0	0	0
Totals × λ_3	0	0	−6	+12	6

[a]The number of observations upon which the total is based.

sum of all scores for the study. The next three values reassure us that we have met the requirement that contrast weights must sum to zero. The final three values (24, 0, and 6) are the sums of $T\lambda$ products, or the L's we require to compute contrasts. Since SS's are given by $L^2/n\Sigma\lambda^2$, our three contrast SS's are:

$$\underset{\text{(Nonhosp. vs. Hosp.)}}{\text{Contrast } SS_1} = \frac{(24)^2}{30(4)} = 4.8$$

$$\underset{\text{(Psy. vs. Comp.)}}{\text{Contrast } SS_2} = \frac{(0)^2}{30(2)} = 0.0$$

$$\underset{\text{(Trad. vs. Mil.)}}{\text{Contrast } SS_3} = \frac{(6)^2}{30(2)} = 0.6$$

Adding these three SS's yields 5.4 which is equal to the total between-treatments SS based on 3 df, as shown in Table 3.2. Special note should be made of the fact that n is always the number of cases upon which the total (T) is based. In the present contrasts, each T is based on 30 observations (10 at each of three age levels). Had the number of observations in the $4 \times 3 = 12$ conditions been unequal, we would have redefined T for each of the 12 conditions as the mean score for that condition multiplied by the harmonic mean of the n's of the 12 conditions. In that case the treatment sums would have become the sum of the three redefined values of T, since each treatment total is made up of three conditions, one at each age level. The n for that treatment total would then become three times the harmonic mean n. We postpone interpretation of our contrasts until we come to our final table of variance.

We turn next to the decomposition of the patients' age factor into two orthogonal contrasts. In this case, since age is an ordered variable (rather than categorical or nominal) we turn to the table of orthogonal polynomials (Table A.1) in the Appendix to find the weights for a linear and quadratic trend. We see that for a three-level factor the weights for linear and quadratic trends are −1, 0, +1 and +1, −2, +1, respectively. The linear trend addresses the question of whether older patients benefit more (or less) from the average treatment than younger patients. The quadratic trend addresses the question of whether middle-aged patients benefit more (or less) from the average treatment than do (the average of) the older and younger patients. Table 3.4 shows the age effect totals of our earlier table, the weights assigned for each of our two orthogonal polynomial contrasts, and the product of the totals × the weights ($T\lambda$).

We compute the contrast SS for the linear trend by

$$\frac{L^2}{n\Sigma\lambda^2} = \frac{[16(+1) + 28(0) + 16(-1)]^2}{40[(+1)^2 + (0)^2 + (-1)^2]} = \frac{(0)^2}{40(2)} = 0.0$$

and we compute the contrast SS for the quadratic trend by

$$\frac{L^2}{n\Sigma\lambda^2} = \frac{[16(+1) + 28(-2) + 16(+1)]^2}{40[(+1)^2 + (-2)^2 + (+1)^2]} = \frac{(-24)^2}{40(6)} = 2.4$$

Table 3.4. *Age effect totals and contrast weights*

Patient age	Totals	Linear (λ_1)	Quadratic (λ_2)	Totals × λ_1	Totals × λ_2
Old	16 (40[a])	+1	+1	16	16
Middle age	28 (40[a])	0	−2	0	−56
Young	16 (40[a])	−1	+1	−16	16
Σ	60 (120[a])	0	0	0	−24

[a]The number of observations upon which the total is based.

Adding these two *SS* yields 2.4, which is equal to the total between-age *SS* based on 2 *df* and shown in our table of variance (Table 3.2). We postpone interpretation of our contrasts until we come to our final table of variance.

Crossing contrasts

Having decomposed the main effects of treatments and of age, we turn our attention now to the decomposition of the interaction effects. Many kinds of contrasts could be employed to decompose the interaction. One that is frequently used in two-factor studies addresses the question of whether the contrasts of one main effect are altered (modified or moderated) by the contrasts of the other main effect. For example, we might ask whether the predicted effect of hospitalization varies as a function of linear trend in age, for example, so that younger patients benefit more than older ones from nonhospitalization. Constructing the contrast weights for these crossed (interaction) contrasts is accomplished by multiplying the contrast weight (λ) defining the column effect contrast (treatments in this example) by the contrast weight (λ) defining the row effect contrast (patient age in this example), as shown in Table 3.5.

Table 3.5. *Construction of crossed (interaction) contrast weights by multiplying column × row contrast weights*

			Treatments				
			Nonhospitalization		Hospitalization		
Patient age	Age contrast	Treatment contrast:	Psychoth. +1	Companion. +1	Tradit. −1	Milieu −1	Σ[a]
Old	+1		+1	+1	−1	−1	0
Middle	0		0	0	0	0	
Young	−1		−1	−1	+1	+1	0
		Σ[a]	0	0	0	0	0

[a]Note that the row and column sums must be zero for contrast weights that are exclusively part of interaction effects.

To obtain the entries of column 1, we multiply the heading λ of column 1 (psychotherapy) by each of the three row λ's (patient age). In this case, multiplying $+1$ by $+1$, 0, -1 in turn yields $+1$, 0, -1. Results for column 2 (companionship) are identical. For column 3 (traditional), we multiply -1 by $+1$, 0, -1 and obtain -1, 0, $+1$; the same result is obtained for column 4 (milieu).

Since we had three treatment effect contrasts and two age effect contrasts, we can obtain six crossed-interaction contrasts by crossing each treatment contrast by each age contrast. These six contrasts are:

(1) Hospitalization vs. nonhospitalization × Linear trend in age
(2) Psychotherapy vs. companionship × Linear trend in age
(3) Traditional vs. milieu therapy × Linear trend in age
(4) Hospitalization vs. nonhospitalization × Quadratic trend in age
(5) Psychotherapy vs. companionship × Quadratic trend in age
(6) Traditional vs. milieu therapy × Quadratic trend in age

Table 3.6 shows the construction of the weights for these crossed (interaction) contrasts, and Table 3.7 shows the computation of the sums of squares for two of these contrasts; numbers (1) and (4) above. The sums of squares of the remaining contrasts, which were similarly computed, are reported in Table 3.8.

Table 3.8, which repeats the two-factor analysis of variance (that was shown earlier in Table 3.2), is now fully decomposed so that we have a contrast for each of the three *df* for treatments, for each of the two *df* for age levels, and for each of the six *df* for the treatments × age levels interactions. The last column gives the magnitudes of each effect in terms of *r*, and we shall interpret our results in order of their effect size. It should be noted that to learn the direction of each of these effects we must examine the means and/or the residuals defining main effects and interactions. The table of variance does not yield that information.

(H) *Hospitalization:* Better outcomes accrue to the nonhospitalized than to the hospitalized.
(Q) *Quadratic trend:* Better outcomes accrue to the middle-aged, than to the average of the young and old.

Table 3.6 *Construction of crossed (interaction) contrast weights by multiplying column and row contrast weights: Six contrasts*

			Treatments			
			Nonhospitalization		Hospitalization	
Patient age	Age contrast		Psychoth.	Companion.	Tradit.	Milieu
		Treatment contrast (1)	+1	+1	−1	−1
Old	+1		+1	+1	−1	−1
Middle	0		0	0	0	0
Young	−1		−1	−1	+1	+1
		Treatment contrast (2)	+1	−1	0	0
Old	+1		+1	−1	0	0
Middle	0		0	0	0	0
Young	−1		−1	+1	0	0

		Treatment contrast (3)	0	0	−1	+1
Old	+1		0	0	−1	+1
Middle	0		0	0	0	0
Young	−1		0	0	+1	−1
		Treatment contrast (4)	+1	+1	−1	−1
Old	+1		+1	+1	−1	−1
Middle	−2		−2	−2	+2	+2
Young	+1		+1	+1	−1	−1
		Treatment contrast (5)	+1	−1	0	0
Old	+1		+1	−1	0	0
Middle	−2		−2	+2	0	0
Young	+1		+1	−1	0	0
		Treatment contrast (6)	0	0	−1	+1
Old	+1		0	0	−1	+1
Middle	−2		0	0	+2	−2
Young	+1		0	0	−1	+1

Table 3.7. *Computation of sums of squares for contrasts (1) and (4) of Table 3.6*

a. Subtable of condition totals from Table 3.1 (n = 10 per condition)

	Nonhospitalization		Hospitalization	
Patient age	Psychotherapy	Companionship	Traditional	Milieu
Old	5	7	2	2
Middle	9	7	4	8
Young	7	7	0	2

b. Subtable of residuals defining interaction[a]

	Nonhospitalization		Hospitalization	
Patient age	Psychotherapy	Companionship	Traditional	Milieu
Old	−1	1	1	−1
Middle	0	−2	0	2
Young	1	1	−1	−1

c. Subtable of contrast computations

$$\text{Contrast } SS = \frac{L^2}{n\Sigma\lambda^2}$$

$$\text{Contrast } SS\ (1) = \frac{[-1(+1) + 1(+1) + 1(-1) + -1(-1) + 0(0) + -2(0) + 0(0) + 2(0) + 1(-1) + 1(-1) + -1(+1) + -1(+1)]^2}{10[(+1)^2 + (+1)^2 + (-1)^2 + (-1)^2 + (0)^2 + (0)^2 + (0)^2 + (0)^2 + (-1)^2 + (-1)^2 + (+1)^2 + (+1)^2]} = \frac{(-4)^2}{10(8)} = 0.2$$

$$\text{Contrast } SS\ (4) = \frac{[-1(+1) + 1(+1) + 1(-1) + -1(-1) + 0(-2) + -2(-2) + 0(+2) + 2(+2) + 1(+1) + 1(+1) + -1(-1) + -1(-1)]^2}{10[(+1)^2 + (+1)^2 + (-1)^2 + (-1)^2 + (-2)^2 + (-2)^2 + (+2)^2 + (+2)^2 + (+1)^2 + (+1)^2 + (-1)^2 + (-1)^2]} = \frac{(12)^2}{10(24)} = 0.6$$

[a]Computed from subtable (a) by subtracting the grand mean, column effects, and row effects. See our discussion of interaction effects in Chapter 1. Each of the residuals is still a sum of 10 observations, but after subtraction of row, column, and grand mean (total) effects. For comprehensive discussion of the computation of interaction effects, see Rosenthal and Rosnow (1984).

Table 3.8. *Two-factor analysis of variance decomposed such that all between condition* dfs *are specified*

Source	*SS*	*df*	*MS*	*F*	*p*	*r*
Treatment	(5.4)	(3)	(1.8)	(36)	(.001)	–
Hospitalized vs. not (H)	4.8	1	4.8	96	.001	.69
Psychother. vs. companion. (P)	0.0	1	0.0	00	–	.00
Tradit. vs. milieu (T)	0.6	1	0.6	12	.001	.32
Age	(2.4)	(2)	(1.2)	(24)	(.001)	–
Linear trend (L)	0.0	1	0.0	0	–	.00
Quadratic trend (Q)	2.4	1	2.4	48	.001	.55
Treatment × age	(1.6)	(6)	(0.267)	(5.34)	(.001)	–
HL	0.2	1	0.2	4	.05	.19
PL	0.1	1	0.1	2	.20	.13
TL	0.1	1	0.1	2	.20	.13
HQ	0.6	1	0.6	12	.001	.32
PQ	0.3	1	0.3	6	.02	.23
TQ	0.3	1	0.3	6	.02	.23
Within	5.4	108	.05			

(T) *Traditional:* If hospitalized, better outcomes accrue to those receiving milieu therapy rather than traditional treatment.

(HQ) *Hosp. × Quad.:* Hospitalization is relatively more effective for the middle-aged than for the young and old (averaged) but nonhospitalization is relatively more effective for the young and old (averaged).

(PQ) *Psych. × Quad.:* If not hospitalized, psychotherapy is more effective for the middle-aged than for the young and old (averaged) but the companionship program is relatively more effective for the young and old (averaged).

Interpretation of the remaining contrasts (including TQ, which is as large as PQ) is left to the reader. Also left to the reader is the question: How might the original data have been constructed to illustrate the desired effects? The final section of this chapter provides an example that may deepen the reader's understanding of the nature of contrasts.

It will be noted that the interpretation of contrasts based on main effects is considerably easier than the interpretation of contrasts based on interaction effects (crossed contrasts in our example). When we consider the quadratic trend in age level, for example, we have only to look at the row totals (or means if the *n*'s were unequal) to see immediately that the benefit scores of the middle-aged are higher than those of the average of the young and old (e.g., see Table 3.1). However, inspection does not so immediately reveal the nature of the crossed contrasts. For example, when we consider the crossed contrast of Hospitalization × Quadratic trend in age (which we know from Table 3.8 to be significant statistically and of at least moderate size), what is it that we should look for in Table 3.1 to tell us what it means?

We cannot use the condition totals directly to tell us about this interaction contrast or any other interaction contrast. That is because the condition totals (or means) are made up partially of the grand mean, partially of the column effects, partially of the row effects, and partially of the interaction effects. In Chapter 1, our discussion of interaction effects gives more detail on this, but here it is enough to note that if we want a clear look at an interaction effect we must remove the column and row effects; removal of the grand mean is optional for purposes of interpretation. Table 3.1 shows the column totals we want to remove before the interaction can be interpreted. These totals are 21, 21, 6, and 12. The mean of these column totals is 15 so we can re-express the column totals as residuals or deviations from this mean (i.e., as 6, 6, −9, and −3).

"Removing" column effects means setting our residual column effects equal to zero. We can do this by subtracting 6 units from the first two columns (2 from each of the three conditions of those columns), adding 9 units to the third column (3 to each of the three conditions of that column), and adding 3 units to the fourth column (1 to each of the three conditions of that column). When we do this, we change the entries of the original table to those of Table 3.9.

The row totals we want to remove next are shown in Table 3.9 as 16, 28, and 16. The mean of these row totals is 20, so the row totals are re-expressed as deviations or residuals from this mean

Table 3.9. *Data of Table 3.1 after removal of column effects*

	Nonhospitalization		Hospitalization		
Patient age	Psychotherapy	Companionship	Traditional	Milieu	Σ
Old	3	5	5	3	16
Middle	7	5	7	9	28
Young	5	5	3	3	16
Σ	15	15	15	15	60

as −4, +8, and −4. To remove the row effects we add 4 units to the first and third rows (1 to each of the four conditions of that row) and subtract 8 units from the second row (2 from each of the four conditions of that row). When we have done this, Table 3.9 is transformed into Table 3.10

The Hospitalization × Quadratic interaction contrast tells us that hospitalization effects differ for the middle-aged versus the average of the young and old. Another way to phrase this is to say that the quadratic trend in age differs as a function of the hospitalization condition. For the present data we can now see that the middle-aged are relatively better off hospitalized (compared to the young and old averaged) and relatively worse off not hospitalized (compared to the young and old averaged). Because the weights for this contrast are the same for columns 1 and 2 and for columns 3 and 4, we can simplify Table 3.10 by combining the conditions for which the contrast weights are identical. This is done in Table 3.11.

Table 3.10. *Data of Table 3.9 after removal of row effects*

	Nonhospitalization		Hospitalization		
Patient age	Psychotherapy	Companionship	Traditional	Milieu	Σ
Old	4	6	6	4	20
Middle	5	3	5	7	20
Young	6	6	4	4	20
Σ	15	15	15	15	60

Table 3.11. *Data of Table 3.10 combining columns of identical contrast weights*

Patient age	Nonhospitalization	Hospitalization	Σ
Old	10	10	20
Middle	8	12	20
Young	12	8	20
Σ	30	30	60

Table 3.11 shows clearly that middle-aged patients are relatively better off hospitalized while the young and old (averaged) are relatively better off not hospitalized. It also shows that hospitalization and nonhospitalization lead to oppositely directioned linear trends (the Hospitalization × Linear trend contrast of earlier tables). Younger patients do worse when hospitalized, better when not hospitalized, a contrast that was significant at $p < .05$, $F(1,108) = 4.00$, $r = .19$.

Constructing data tables: A procedure for teaching and learning

One way to have our students demonstrate to themselves that they understand contrasts is to be able to generate them in the rows, columns, and interactions of a table of data that they create from scratch. Generating data tables to illustrate particular effects is basically the reverse of the process of finding the residuals defining the interaction. Whereas this process of finding the residuals (sometimes called "mean polish") involves subtracting the grand mean, row effects, column effects, and so forth, in order to be able to examine the residuals, constructing a data table involves *adding* the grand mean, row effects, column effects, and so forth, in order to be able to create the desired observed data. For example, the data of our original table (Table 3.1) could be constructed as follows:

1. Select an average cell value

Having decided to construct a design of 12 conditions arranged as a four treatment by three age levels matrix, we begin by assigning some average cell value for each cell (in this case, the value 5). This first step is shown in Table 3.12.

Table 3.12. *Specimen data showing the same average value for each cell*

Age	Treatments A	B	C	D	Σ
Old	5	5	5	5	20
Middle age	5	5	5	5	20
Young	5	5	5	5	20
Σ	15	15	15	15	60

2. Select a row effect

Assume we want to illustrate a quadratic trend effect in our row totals. Since, for three levels, the weights required are −1, +2, −1, we subtract 1 from every entry of row 1, add 2 to every entry of row 2, and subtract 1 from every entry of row 3. This second step is shown in Table 3.13.

Table 3.13. *Data of Table 3.12 after adding a quadratic row effect*

Age	Treatments A	B	C	D	Σ
Old	4	4	4	4	16
Middle age	7	7	7	7	28
Young	4	4	4	4	16
Σ	15	15	15	15	60

3. Select a column effect

Assume we want to illustrate treatment effects, such that A and B are equal to each other and 3 units greater than D which, in turn, is 2 units greater than C. The weights satisfying those requirements (and adding to zero) are 2, 2, −3, −1 for treatments A, B, C, D, respectively. Therefore, we add 2 to every entry of treatments A and B, subtract 3 from every entry of treatment C, and subtract 1 from every entry of treatment D. This step is shown in Table 3.14.

Table 3.14. *Data of Table 3.13 after adding a column effect such that* $A = B > D > C$

Age	Treatments A	B	C	D	Σ
Old	6	6	1	3	16
Middle age	9	9	4	6	28
Young	6	6	1	3	16
Σ	21	21	6	12	60

4. Select an interaction effect

Suppose we want to illustrate interaction effects such that treatments A and C show linear trends in age that are in opposite directions to each other while treatments B and D show quadratic trends in age that are in opposite directions to each other. The weights representing these effects are shown in Table 3.15.

Table 3.15. *Weights for linear and quadratic elements in an interaction effect*

Age	Treatments A	B	C	D	Σ[a]
Old	−1	+1	+1	−1	0
Middle age	0	−2	0	+2	0
Young	+1	+1	−1	−1	0
Σ[a]	0	0	0	0	0

[a] Note that the row and column sums must be zero for contrast weights that are exclusively part of interaction effects.

Then, adding these interaction effects to the effects we had built up before (grand mean plus row effect plus column effect) yields Table 3.16. Turning back to Table 3.1, we see that these two tables are identical.

Table 3.16. *Data of Table 3.14 after adding the interaction effect of Table 3.15*

	Treatments				
Age	A	B	C	D	Σ
Old	5	7	2	2	16
Middle age	9	7	4	8	28
Young	7	7	0	2	16
Σ	21	21	6	12	60

4
Wired-in, pre-analyses, multiples, and proportions

Wired-in contrasts

Many of our standard experimental designs have their contrasts "wired in"; that is, contrasts are inherent in the designs. In a 2 × 2 factorial design, for example, the column effect, the row effect, and the interaction effect are all contrasts, and with equal *n*'s (or with unweighted means approaches) they are orthogonal contrasts. A 2 × 2 × 2 factorial design is made up of seven wired-in contrasts: a column effect (left vs. right), a row effect (top vs. bottom), a block effect (front vs. back), a column × row, a column × block, a row × block, and a column × row × block effect. More generally, any two-level factor can be seen as a contrast, as can any crossing of two (or more) two-level factors.

For example, in a recently employed 2 × 2 × 2 × 2 × 3 × 5 design, the four main effects with two levels, the six two-way interactions involving only the two-level factors, the four three-way interactions involving only the two-level factors, and the four-way interaction involving only the two-level factors could all be seen as wired-in contrasts (Rosenthal, 1981; Rosenthal, Hall, DiMatteo, Rogers, & Archer, 1979).

These wired-in contrasts are extremely useful and bring considerable power to bear on the questions addressed. They are greatly to be preferred to more diffuse, unfocused sources of variance, that is, those having more than a single *df* and which will not be followed up by more focused questions. However, it sometimes happens that the use of wired-in contrasts misleads us in our attempt to test the statistical significance of our effect and to estimate its effect size.

Table 4.1. *Predicted results of an experiment (Data are sums of benefit scores; n = 10)*

	Type of patient				
	Depressive		Obsessive-compulsive		
	Medication	Control	Medication	Control	Σ
Psychotherapy	8 (10[a])	0	0	0	8 (40[a])
Control	0	0	0	0	0
Σ	8 (20[a])	0	0	0	8 (80[a])

[a]The number of observations upon which the sum is based.

Misleading wired-in contrasts: An example

Consider the following as an example in which contrasts have this weakening effect. Suppose we have hypothesized that depressive patients, but not obsessive-compulsive patients, will benefit from a new form of psychotherapy if and only if a particular medication has also been prescribed. Since we have three factors in our study – (a) type of patient, (b) psychotherapy versus control, and (c) medication versus control – we think of a $2 \times 2 \times 2$ factorial design. Table 4.1 shows the benefit scores our theory might predict for each of the eight conditions of the $2 \times 2 \times 2$ design, and Table 4.2 shows the analysis of variance of these data.

Table 4.2. *Analysis of variance of data of Table 4.1*

Source	*df*	*MS*	*F*	*p*	*r*
Type of patient	1	0.8	1.0	.32	.12
Psychotherapy	1	0.8	1.0	.32	.12
Medication	1	0.8	1.0	.32	.12
Type × psych.	1	0.8	1.0	.32	.12
Type × medic.	1	0.8	1.0	.32	.12
Psych. × medic.	1	0.8	1.0	.32	.12
Type × psych. × medic.	1	0.8	1.0	.32	.12
Within	72	0.8			

Imagine now that the data we predicted, as shown in Table 4.1, were exactly the data we had obtained. Suppose further that we had performed the standard $2 \times 2 \times 2$ analysis of these data. The table of variance reveals that none of our effects, that is, wired-in contrasts, would have been significant and none of our effect sizes would have exceeded an r of .12. The reason is that our $2 \times 2 \times 2$ analysis has dissipated the real question we wanted to ask of our data over seven different questions we really did not want to ask.

The question we really wanted to ask (as shown in Table 4.1) was whether those depressive patients given our new form of psychotherapy *and* a particular medication benefited more than did the patients in any other condition. We construct our contrast weights to test this hypothesis by simply subtracting the *mean* of all the expected values from *each* of the expected values. In the present example the expected values are given in Table 4.1 and their mean is 1.0. When we subtract 1.0 from each of the expected values, we have seven values of -1 and a single value of $+7$ as the eight weights for our contrast.

Computing our contrast then yields

$$SS \text{ contrast} = \frac{L^2}{n\Sigma\lambda^2}$$

$$= \frac{[8(+7) + 0(-1) + 0(-1) + 0(-1) + 0(-1) + 0(-1) + 0(-1) + 0(-1)]^2}{10[(7)^2 + (-1)^2 + (-1)^2 + (-1)^2 + (-1)^2 + (-1)^2 + (-1)^2 + (-1)^2]}$$

$$= \frac{(56)^2}{10(56)} = 5.6$$

which, when divided by the mean square for error (from Table 4.2) yields $F(1,72) = 7.0$, $p = .01$, $r = .30$. Thus, employing the contrast specifically designed to test our hypothesis yields both a significant result and one substantially larger in effect size than did the less direct series of seven contrasts, none of which was really a test of the hypothesis of interest to us.

Pre-analysis of an experiment

A very useful step in the design stages of our studies is to construct a table of the means we expect to obtain in every condition. Then, when we subtract the grand mean of all of our expected mean values from each of our expected means, we have the weights required for the overall contrast testing our theory. If we have arranged our conditions into a two-way or higher-order table, it will also be instructive to compute the analysis of variance directly on our predicted values. This will show us, in terms of the relative magnitudes of the mean squares and effect sizes, how our predictions would be expressed in a standard analysis of variance type format.

As an example, imagine the following experiment on the effects of interpersonal expectancy on subjects' performance (Rosenthal, 1966, 1969, 1976, in press; Rosenthal & Jacobson, 1968; Rosenthal & Rubin, 1978). Subjects who are high or low on sensitivity to nonverbal cues are first selected. Half of each of these two groups of subjects are then assigned to experimenters who have been led to expect good performance, and half are assigned to experimenters who have been led to expect poor performance. Each of these four conditions is further subdivided into a condition in which subjects are made to feel above-average levels of evaluation apprehension (concern over what the experimenter will think of them) and a condition in which subjects are made to feel below-average levels of evaluation apprehension. The

Table 4.3. *Predicted results of an experiment*

	Evaluation apprehension				
	Low		High		
Nonverbal sensitivity	Low expectancy	High expectancy	Low expectancy	High expectancy	Σ
High	2	4	0	6	12
Low	4	2	2	4	12
Σ	6	6	2	10	24

Table 4.4. *Analysis of variance of predicted results in Table 4.3*

Source	*df*	*MS*
Expectancy	1	8.0
Evaluation apprehension	1	0.0
Sensitivity	1	0.0
Expect. × eval. app.	1	8.0
Expect. × sensitivity	1	8.0
Eval. app. × sensitivity	1	0.0
Exp. × eval. app. × sens.	1	0.0

theoretical reasoning behind this design would be that subjects' performance levels (the dependent variable) will be more in accord with the experimenter's expectancies when the subjects are more sensitive to nonverbal cues or are feeling more anxious about being evaluated by the experimenter, or both. Table 4.3 shows the hypothesized results (not the actual results) of this study according to the investigator's theory, and Table 4.4 shows the results of the 2 × 2 × 2 analysis of variance on the predicted values.

Table 4.4 reveals that the overall prediction pattern of the preceding table (which would have been a set of contrast weights if we had subtracted the grand mean prediction of $24/8 = 3.0$ from each prediction) has three components: (a) a main effect of expectancy, (b) an interaction of expectancy and evaluation apprehension, and (c) an interaction of expectancy and sensitivity to nonverbal cues. Since in this particular study we might be interested in the results of the overall contrast as well as a comparison of the relative magnitudes of the three effects predicted for the analysis of variance, we would probably compute both.

Inspection of our predicted results and construction of the appropriate two-way tables shows us that the expectancy effects we hypothesized were predicted to be greater under (a) conditions of high (rather than low) evaluation apprehension and (b) conditions of high (rather than low) sensitivity to nonverbal cues.

Multiple planned and unplanned contrasts

So far our discussion of contrasts has focused on the situation in which we have planned each contrast. When each contrast has been planned, we need not generally be concerned about the significance of any "overall" F tests. The analysis of variance with multilevel factors serves in these cases not as a source of multi-*df* (for the numerator) F tests, but as an efficient way to organize the various sources of variance and especially the sources serving as our error term or terms.

If we have planned an unusually large number of contrasts for our investigation, however, and want to take into account the fact that as the number of planned contrasts increases the probability of a type I error also increases, we might want to adjust our alpha (significance) level accordingly. A simple and generally conservative way to make this adjustment is based on the *Bonferroni approach* (Hays, 1981; Morrison, 1976; Rosenthal & Rubin, 1983; 1984; Snedecor & Cochran, 1980). We need only divide our favorite overall alpha level, say .05, by the number of contrasts we have planned, to find the adjusted alpha level we now require for each contrast. Thus, if we set alpha for the set of contrasts to .05 and we plan eight contrasts, we could adjust our alpha per contrast to $.05/8 = .00625$.

Since the Bonferroni approach does not require that we set the same alpha level for each contrast, we can allocate the total alpha of .05 unequally. For example, if we want greater power for four of our contrasts, and less for the remaining four, we could set the former four alphas at $p = .01$ and the latter four at $[.05 - 4(.01)]/4 = .0025$. Our eight contrasts are now tested at .01, .01, .01, .01, .0025, .0025, .0025, .0025, respectively, which when added up, yield .05, our overall contrast-set alpha level (Harris, 1975; Myers, 1979; Rosenthal & Rubin, 1983; 1984). We should note, however, that these procedures require that we have planned the contrasts before any results have been made available to us.

If all the planned contrasts fall under the umbrella of an appropriate overall F (as is the case in one-way analyses of variance), and if the overall F is significant, we receive sufficient protection for our multiple contrasts so that the Bonferroni procedure is not required. If the overall F is not significant, the Bonferroni procedure may be employed but is less and less indicated as the number of planned contrasts decreases and the strength of the theoretical and empirical basis for each contrast increases.

If the contrasts being tested were not planned, the Bonferroni procedure can still be employed but only if we are able to specify the total number of contrasts implicitly examined in the process of finding the contrast we want to test for significance.[1] As an alternative to the Bonferroni approach the overall F approach can be employed. Thus, if the appropriate overall F is significant, contrasts found significant under the umbrella of the appropriate overall F are sufficiently protected on the average (Balaam, 1963; Snedecor & Cochran, 1967). If (a) the appropriate overall F is not significant and (b) the Bonferroni approach cannot be employed because the total number of contrasts implicitly examined cannot be specified, and (c) we have formed our contrast simply on the basis that it "looked like a big effect," we can still do a significance test with a meaningful alpha level, the Scheffé test (Hays, 1981; Winer, 1971). The Scheffé test, however, should not be employed as additional protection if the appropriate overall F test was significant, since that is counter to the spirit and logic of the Scheffé test. It puts the contrast in double jeopardy and is analogous to sending out a rowboat to determine whether the oceans are calm enough for an ocean liner, as Box has said of a related issue (Box, 1953).

If a Scheffé test is desired, it is easily accomplished by adjusting the value of F required to reach any particular level of significance. Suppose we have an experiment comparing six treatment conditions, and we have assigned six patients to each condition. Assuming our level of α to be .05, it would take an overall $F(5,30) = 2.53$ to reach significance at .05. To perform the Scheffé test we multiply this overall critical value of F by the number of df for the numerator to yield a new critical value of F that must be reached before we can safely call significant at $p < .05$ a contrast chosen solely because it "looked big." In our example, the

[1] For example, if we have 8 means in a study and we want to test the largest difference for significance, that implies we have implicitly compared all possible pairs of means, so we have made $[k(k - 1)]/2 = [8 \times 7]/2 = 28$ possible comparisons.

df for the numerator = 5 so our Scheffé-required F is

$$\text{Scheffé } F = \text{Ordinary overall } F \times df \text{ numerator} = 2.53 \times 5 = 12.65$$

In this section when we have referred to overall F tests we have emphasized that they be "appropriate." We have done this because both in scientific and in statistics texts there is some confusion over what constitutes an "overall" F test. There is no confusion in a one-way analysis of variance since there is only one F to be computed. However, imagine a two-factor analysis of variance with five rows and six columns. The F for rows (with 4 *df* in the numerator) is the appropriate overall F for contrasts among the row totals (or means). The F for columns (with 5 *df* in the numerator) is the appropriate overall F for contrasts among the column totals (or means). The F for interaction (with 20 *df* in the numerator) is the appropriate overall F for contrasts among the 30 residuals of the design, that is, the means of the 30 conditions *after* removal of the row and column effects and, optionally, the grand mean. The F for interaction is *not* the appropriate overall F for contrasts among the 30 means or totals of the 30 conditions. If contrasts are to be made among the 30 means or totals, and an overall F is required, a new F must be computed which will have 29 *df* in the numerator. This new F will be the equivalent of the one-way analysis of variance F computed among all 30 conditions. In a two-dimensional factorial analysis with a single error term, the required F overall can be computed by

$$F \text{ overall} = \frac{(F_{\text{Row}} \times df_{\text{Row}}) + (F_{\text{Column}} \times df_{\text{Column}}) + (F_{\text{R}\times\text{C}} \times df_{\text{R}\times\text{C}})}{df_{\text{Row}} + df_{\text{Column}} + df_{\text{R}\times\text{C}}}$$

Contrasts in proportions

Our discussion of contrasts has been primarily within the context of the analysis of variance. However, as noted in Chapter 1, the use of contrasts is not restricted to that context. In this section we show the use of contrasts in examining the data of contingency tables in which there are two rows and three or more columns. In these situations the data are frequency counts in which we form some substantively meaningful proportion for each column (by dividing one of the cells of the array by the sum of the counts found in that column). The following table illustrates this usage. The data were made available by Smadar Levin (1974) from her study of men and women listed in *Who's Who in America.* For her 54 career men and 40 career women she recorded the number of children they had and cast the counts into the format shown in Table 4.5. Her hypothesis was that as the number of children increased from zero to three or more, career women would form a smaller proportion of the total of career people found at that level.

The first two rows of Table 4.5 (labeled A and B) present the raw counts; the third row presents the totals for each column (A+ B, labeled N); and the fourth row presents the proportion (P) of the persons in that column who are women. The fifth row gives the squared standard error of the proportion for each column. The last row shows the contrast weights appropriate to the test of Levin's prediction. It is, of course, the set of linear trend weights for four conditions that we have seen before (see Table A.1 in the Appendix.)

The significance test for the contrast in proportions is not an F but a standard normal deviate, Z. The associated p levels are found in Table A.2 in the Appendix. The formula, suggested by Donald B. Rubin in a personal communication (1981), is:

$$z = \frac{\Sigma(P\lambda)}{\sqrt{\Sigma(S_P^2\,\lambda^2)}}$$

Table 4.5. *Number of career men and women found at four levels of number of children (after Levin, 1974)*

	Number of children				
	0	1	2	3+	Σ
(A) Men	6	10	15	23	54
(B) Women	17	7	9	7	40
(N) = (A + B)	23	17	24	30	94
(P) = B/N	.74	.41	.38	.23	
$S_P^2 = \dfrac{P(1-P)}{N}$	.0084	.0142	.0098	.0059	
λ	+3	+1	−1	−3	

where $P = B/N$ the proportions of interest

and $S_P^2 = \dfrac{P(1 - P)}{N}$

Applied to the data above we find

$$z = \frac{\Sigma P\lambda}{\sqrt{\Sigma S_P^2 \lambda^2}}$$

$$= \frac{.74(+3) + .41(+1) + .38(-1) + .23(-3)}{\sqrt{.0084(9) + .0142(1) + .0098(1) + .0059(9)}}$$

$$= \frac{1.56}{\sqrt{.1527}} = 4.00$$

$p = .00003$, one-tailed. Thus, Levin's hypothesis was strongly supported by this test of significance.

The size of the effect can be judged from direct inspection of the run of proportions from .74 all the way to .23. Incidentally, the χ^2 (1) employing only the zero versus three or more children columns is found to be 13.44, $Z = 3.67$, $p = .00012$.

Other procedures for testing linear trends in proportions are available, but because they are more complex computationally and cannot be directly applied to contrasts other than linear trends they are not described here (Snedecor & Cochran, 1967, p. 247). One flexible procedure that can be employed when proportions are not too extreme, say between .15 and .85, is to give each of the persons in the study a score of zero or one on the two-level variable (e.g., 1 = female, 0 = male) and proceed with the standard analysis of variance methods we have described elsewhere (Rosenthal & Rosnow, 1984). The appropriateness of this procedure is well documented (e.g., Cochran, 1950; D'Agostino, 1971; Edwards, 1972; Lunney, 1970; Rosenthal & Rosnow, 1984; Snedecor & Cochran, 1967; Winer, 1971).

5

Repeated-measures designs

Within subjects

So far we have considered contrasts only in the situation of "between-subjects" designs. That is, every element to which we assigned a weight, prediction or λ was comprised of different sampling units. But it happens frequently that we measure our sampling units more than once, as in studies of learning, or in diachronic analyses (analyses over time to measure long-term changes) of planned interventions, maturation, and the like. Indeed, sound principles of experimental design lead us frequently to employ "within-subject" designs, or repeated-measures designs, in order to increase experimental precision and, therefore, statistical power. Our need for employing contrasts is just as great within repeated-measures designs as in between-subjects designs. Consider the following study: four subjects each are measured on three occasions one week apart on the same test of verbal skill. The entries in Table 5.1 and Figure 5.1 are the observations of

Table 5.1. *Results of a study of verbal skill measured on three occasions for each of four subjects*

	Occasions			
Subjects	1	2	3	Σ
A	8	12	10	30
B	7	11	9	27
C	3	5	1	9
D	2	4	0	6
Σ	20	32	20	72

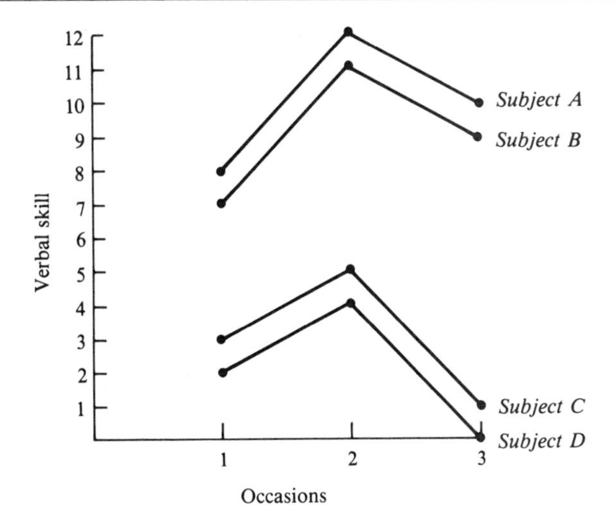

Figure 5.1. Results of a study of verbal skill (Table 5.1).

verbal skill scores obtained on three occasions by each of the four subjects.

The preliminary analysis in Table 5.2 tells us that verbal skill performance varies from occasion to occasion. However, that was not our question. Our primary concern was whether there was a steady increase in improvement over time – that is, a linear trend with weights of $-1, 0, +1$. Our secondary question was whether there was a tendency for scores first to improve and then to fall back to the original level – that is, a quadratic trend with weights of $-1, +2, -1$.

Table 5.2. *Preliminary analysis of variance of data in Table 5.1*

Source	*SS*	*df*	*MS*	*F*	*p*
Between subjects	150	3	50.00	–	–
Within subjects	(32)	(8)			
Occasions	24	2	12.00	9.00	.02
Occasions × subjects	8	6	1.33		

Computing the contrast analysis

Accordingly, we simply apply our standard procedure for computing contrasts to the column totals of our table of results:

	Occasions (Sums of four scores)		
Column Totals:	20	32	20
Linear λ:	−1	0	+1
Quadratic λ:	−1	+2	−1

$$\frac{L^2}{n\Sigma\lambda^2} \text{ for linear trend } = \frac{(-20 + 0 + 20)^2}{4(2)} = 0$$

$$\frac{L^2}{n\Sigma\lambda^2} \text{ for quadratic trend } = \frac{(-20 + 64 - 20)^2}{4(6)} = \frac{(24)^2}{24} = 24$$

These two contrasts show that the total variation among the column totals was entirely due to the quadratic trend. We can now rewrite the table of variance subdividing the occasions effect into the two contrast effects. These new results are shown in Table 5.3. Each of our contrasts was tested against the same error term against which we tested the overall effect of occasions. Under most circumstances, that is probably the wisest choice of error term for each contrast. It is not, however, the only choice we have. We can also construct a specific error term for each contrast. Just as the error term for the occasions effect is the crossing of the occasions effect by the subjects effect, so too is the unique error term for each contrast constructed by crossing that contrast by

Table 5.3. *Rewritten analysis of variance of data in Table 5.1*

Source	*SS*	*df*	*MS*	*F*	*p*
Between subjects	150	3	50.00	–	–
Within subjects	(32)	(8)			
Occasions	24	2	12.00	9.00	.02
Linear trend	0	1	0	–	–
Quadratic trend	24	1	24.00	18.00	.006
Occasions × subjects	8	6	1.33		

Table 5.4. *Original data corrected for grand mean*

Subjects	Occasions 1	2	3	Σ	$\bar{X}$
A	2	6	4	12	4
B	1	5	3	9	3
C	−3	−1	−5	−9	−3
D	−4	−2	−6	−12	−4
Σ	−4	8	−4	0	
$\bar{X}$	−1	2	−1		0

the subject effect. Practically speaking, we are likely to prefer the use of these unique error terms for significance testing and effect size estimation only when the various unique error terms differ dramatically from each other and from the pooled or overall error term.

It will be best to begin by having a look at the overall interaction of occasions × subjects, which is the set of residuals remaining after we have removed the grand mean, row effects, and column effects. After subtracting the grand mean from the original data shown in Table 5.1, we obtain the corrected data shown in Table 5.4. Next we subtract off the row effect from each entry of each row; this gives us the data shown in Table 5.5. We now subtract the column effects, so that only the interaction effects (or

Table 5.5. *Original data corrected for grand mean and row effects*

Subjects	Occasions 1	2	3	Σ	$\bar{X}$
A	−2	2	0	0	0
B	−2	2	0	0	0
C	0	2	−2	0	0
D	0	2	−2	0	0
Σ	−4	8	−4	0	
$\bar{X}$	−1	2	−1		0

residuals) remain; this correction gives us the results in Table 5.6. This table shows the residuals defining the occasions × subjects interaction. The sum of squares for these residuals is easily computed as $\sum^{N=12} X^2 = 8$, the value obtained earlier for our table of variance (Table 5.2) by the usual method of subtraction (i.e., Total *SS* minus Subjects *SS* minus Occasions *SS*). This interaction sum of squares must now be separated into two independent portions, one representing the linear trend × subjects interaction and the other representing the quadratic trend × subjects interaction.

We proceed by computing for each subject a standardized "score" for each contrast. This *standardized contrast score* is defined as $L^2/\Sigma\lambda^2$ for each subject, where L is the sum of each of the subject's residual scores after it has been multiplied by the appropriate λ or contrast weight. Thus, for subject A, her scores (residuals) of -1, 0, $+1$ are multiplied by the linear weights of -1, 0, $+1$ respectively, to form the L for the linear contrast; they are multiplied by -1, $+2$, -1 to form the L for the quadratic contrast. Subject A's L score for the linear contrast, therefore, is $(-1)(-1) + (0)(0) + (1)(1) = 2$. Her standardized contrast score is $L^2/\Sigma\lambda^2 = 2^2/2 = 2$. Subject A's L score for the quadratic contrast is $(-1)(-1) + (0)(+2) + (1)(-1) = 0$. Her standardized contrast score is $L^2/\Sigma\lambda^2 = 0^2/6 = 0$. Table 5.7 shows the L (raw contrast score) and $L^2/\Sigma\lambda^2$ (standardized contrast score) for each subject for both the linear and quadratic contrasts.

The sum of squares for the linear trend × subjects interaction is simply the sum of all of the subjects' standardized contrast

Table 5.6. *Original data corrected for grand mean, row effects, and column effects*

	Occasions			
Subjects	1	2	3	Σ
A	−1	0	1	0
B	−1	0	1	0
C	1	0	−1	0
D	1	0	−1	0
Σ	0	0	0	0

Table 5.7. *Raw and standardized contrast scores for four subjects*

	Linear contrast		Quadratic contrast	
Subject	L	$L^2/\Sigma\lambda^2$	L	$L^2/\Sigma\lambda^2$
A	2	2	0	0
B	2	2	0	0
C	−2	2	0	0
D	−2	2	0	0
Σ	0	8	0	0

scores. In this case, that sum of squares equals 8. The analogous sum of squares for the quadratic trend × subjects interaction is equal to zero. The sum of these two sums of squares is 8, which is just what the table of residuals requires. That is, the linear × subjects plus the quadratic × subjects sums of squares, each with 3 *df*, must be equal to the overall occasions × subjects interaction with 6 *df*. Table 5.8 shows the final table of variance.

This table shows that all of the variation between occasions is due to the quadratic trend, while none of the variation in the

Table 5.8. *Final table of variance with decomposed error terms*

Source	*SS*	*df*	*MS*	*F*	*p*
Between subjects	150	3	50.00	–	–
Within subjects	(32)	(8)			
Occasions	24	2	12.00	9.00	.02
[a]Occasions × subjects	8	6	1.33		
Linear trend	0	1	0	–	–
[a]Linear × subjects	8	3	2.67		
Quadratic trend	24	1	24.00	∞	0
[a]Quadratic × subjects	0	3	0		

[a]Appropriate error term for the just preceding effect.

interaction between occasions and subjects is due to the quadratic trend × subjects component of the total interaction. This yields an unusually large F for the quadratic effect, one tending toward infinity (∞). It seems better practice normally to pool these specialized error terms, especially when each is based on few *df*. Therefore, in the present example we might prefer to compute $F(1,6)$ for quadratic trend as $24/1.33 = 18$, $p = .006$, rather than as $24/0$.

Contrast scores for individuals

So far in our consideration of contrasts on repeated measures we have been interested only in contrasts on groups of subjects; for example, the large quadratic trend effect shown in our original data table applied to the sum of the four subjects measured on three occasions. Sometimes, however, we would like to be able to say for each subject (or whatever the sampling unit might be, for example, groups, countries, school systems, etc.) the degree to which that subject shows any particular contrast. Such would be the case if we wanted to correlate some other attribute (e.g., age, sex, personality score, ability score, etc.) with the degree to which the planned contrast characterized each subject.

A few pages back we encountered the procedure of computing L scores as a step in computing standardized contrast scores in the process of computing specific error terms for contrasts in repeated measures. In that situation, since we were trying to decompose an occasions × subjects interaction, we operated only on the residual scores remaining to each subject after we had subtracted off the grand mean, row effects, and column effects. Now, however, since we want to compute contrast scores in the column effects (occasions), we want to operate on the original data. The results we obtain will be the same whether we subtract from the original data the grand mean and row effects (subject effects) or not, a fact that is characteristic of the analysis of main effects. Thus, we could operate directly on our original data, or on the data from which the grand mean has been removed, or on the data from which the grand mean and the row effects have both been removed. We compute the raw contrast score L by multiplying each of the subject's scores by the appropriate λ or contrast weight. Hence, for subject A of our original data table (see Table 5.1), her scores of 8, 12, 10 are multiplied by the linear weights −1, 0, +1, respectively, to form the L score for the linear contrast. Her scores are multiplied by −1, +2, −1 to form the quadratic contrast score, or by any other weights (summing to zero) representing any other contrast.

Subject A's L score for the linear contrast, based on the data of our first table (original raw scores in Table 5.1), therefore, is $8(-1) + 12(0) + 10(+1) = 2$. Subject A's L score for the quadratic contrast, based on the original data, is $8(-1) + 12(+2) + 10(-1) = 6$. Had we based subject A's linear and quadratic L scores on the "de-meaned" data, these L scores would have been $2(-1) + 6(0) + 4(+1) = 2$, and $2(-1) + 6(+2) + 4(-1) = 6$ respectively, exactly the same as the L scores based on the original data. Had we based subject A's linear and quadratic L scores on the data with both the grand mean and the row effects removed, these L scores would have been $-2(-1) + 2(0) + 0(+1) = 2$, and $-2(-1) + 2(+2) + 0(-1) = 6$, again, exactly the same as the L scores based on the original data.

Once we know how to compute L (contrast) scores for any hypothesis, such as that for linear trend or quadratic trend (or any other set of contrast weights), we can directly test the question of whether the subjects as a group show that particular trend to a significant extent. If the null hypothesis were true, for example, that there were no linear trend in the population, we would expect, on the average, that positive linear trends by some subjects would be offset by negative linear trends by other subjects so that there would be no net linear trend. Thus, under the hypothesis of no linear trend, we expect the mean L (contrast) score for linear trend to be zero. We can, therefore, test the hypothesis of linear trend by a simple t test of the L scores against the value of zero expected if the null hypothesis were true, hence

$$t = \frac{\overline{L} - 0}{\sqrt{S^2/N}}$$

where $\overline{L}$ is the mean of the L scores, S^2 is computed from the L scores, and N is the number of L scores.

For the data in Table 5.1, the L scores are 2, 2, -2, -2, computed as follows:

Subject A	$8(-1) + 12(0) + 10(1) =$	2
Subject B	$7(-1) + 11(0) + 9(1) =$	2
Subject C	$3(-1) + 5(0) + 1(1) =$	-2
Subject D	$2(-1) + 4(0) + 0(1) =$	-2

Therefore, $\bar{L} = 0$, $S^2 = \Sigma(L - \bar{L})^2/(N - 1) = [(2 - 0)^2 + (2 - 0)^2 + (-2 - 0)^2 + (-2 - 0)^2]/3 = 16/3 = 5.33$, $N = 4$, so that

$$t = \frac{0 - 0}{\sqrt{5.33/4}} = \frac{0}{1.15} = 0$$

Since $t^2 = F$, we can equally well compute an F test of the hypothesis of linear trend for these subjects, and since $t = 0$, $t^2 = F = 0$ as well. Examination once more of Table 5.8 reveals that the mean square for linear trend is zero and F is therefore also zero.

As a further illustration of the use of L scores to test hypotheses, we show the t test of the hypothesis of a quadratic trend for the same four subjects. The L scores are obtained as follows:

Subject A	$8(-1) + 12(2) + 10(-1) =$	6
Subject B	$7(-1) + 11(2) + 9(-1) =$	6
Subject C	$3(-1) + 5(2) + 1(-1) =$	6
Subject D	$2(-1) + 4(2) + 0(-1) =$	6

Therefore, $\bar{L} = 6$, $S^2 = \Sigma(L - \bar{L})^2/(N - 1) = [(6 - 6)^2 + (6 - 6)^2 + (6 - 6)^2 + (6 - 6)^2]/3 = 0/3 = 0$, $N = 4$ so that

$$t = \frac{6 - 0}{\sqrt{0/4}} = \infty$$

We note that since $S^2 = 0$, $t = \infty$ and, therefore, since $F = t^2$, F is also ∞ as shown for the quadratic trend effect in Table 5.8.

It will generally be the case that the simple t test of the mean L score against the zero value expected if the null hypothesis were true will yield the same result (in the sense of $t^2 = F$) as the F test of the same contrast when the mean square for the contrast (e.g., linear, quadratic, any other) is tested against the specific error

Table 5.9. *Scores obtained from three subjects on four occasions*

	Occasions				
Subjects	1	2	3	4	Σ
A	−1	+3	−3	+1	0
B	−3	−1	+1	+3	0
C	−1	+1	+1	−1	0
Σ	−5	+3	−1	+3	0

term for that contrast. If, however, we have employed a pooled or aggregated error term for our F test, that will no longer be true, although the results will still tend to be quite similar. We should also note that the S^2 computed for the t test will not generally be the same as the S^2 computed for the F test, but will be larger than the latter by a factor of $\Sigma\lambda^2$. That is, the S^2 computed for the t test differs from the error MS for that particular contrast by a factor of the sum of the squared contrast weights.

We use the data of Table 5.9 and Figure 5.2 to illustrate this

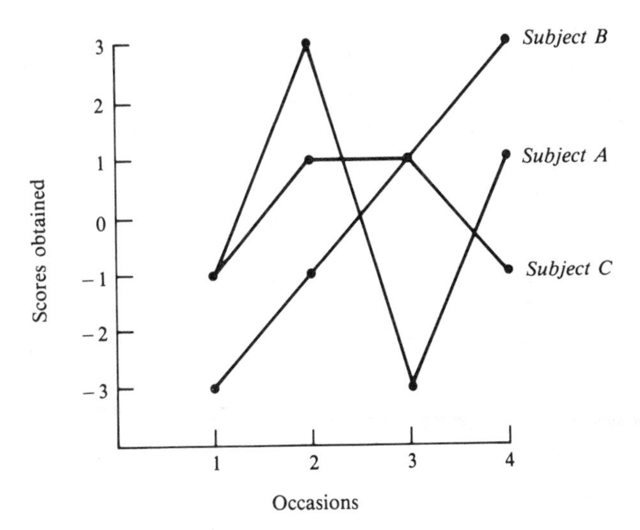

Figure 5.2. Results of a study of three subjects (Table 5.9).

Table 5.10. *Table of variance for the data in Table 5.9*

Source	*SS*	*df*	*MS*	*F*	*p*
Between subjects	0	2	0	–	–
Within subjects	(44)	(9)			
Occasions	14.67	3	4.89	1.00	–
[a]Occasions × subjects	29.33	6	4.89		
Linear trend	6.67	1	6.67	1.00	–
[a]Linear × subjects	13.33	2	6.67		
Quadratic trend	1.33	1	1.33	1.00	–
[a]Quadratic × subjects	2.67	2	1.33		
Cubic trend	6.67	1	6.67	1.00	
[a]Cubic × subjects	13.33	2	6.67		

[a]Appropriate error term for the just preceding effect.

and to serve as a review. Table 5.10 serves as a summary table of variance for the data in Table 5.9.

As a review of how we obtained this table of variance we need the contrast weights or λs for the linear, quadratic, and cubic trends, given that there are four occasions. These weights are -3, -1, $+1$, $+3$; -1, $+1$, $+1$, -1; and -1, $+3$, -3, $+1$ respectively (see Table A.1 in the Appendix). Table 5.11 shows the raw and standardized contrast scores for these three subjects for all three

Table 5.11. *Raw and standardized contrast scores for three subjects based on residuals*

	Linear		Quadratic		Cubic	
Subject	L	$L^2/\Sigma\lambda^2$	L	$L^2/\Sigma\lambda^2$	L	$L^2/\Sigma\lambda^2$
A	−6.67	2.22	−1.33	0.44	13.33	8.89
B	13.33	8.89	−1.33	0.44	−6.67	2.22
C	−6.67	2.22	2.67	1.78	−6.67	2.22
Σ	0	13.33	0	2.67	0	13.33

Table 5.12. *Residuals upon which the scores in Table 5.11 are based*

	Occasions				
Subject	1	2	3	4	Σ
A	0.67	2.00	−2.67	0.00	0
B	−1.33	−2.00	1.33	2.00	0
C	0.67	0.00	1.33	−2.00	0
Σ	0	0	0	0	0

contrasts, and Table 5.12 gives the residuals of the original data of this example.

The scores given in Table 5.11 were obtained by computing the L scores as well as the standardized contrast scores for each subject's raw scores, as shown earlier for the data of our previous example (residuals). The sums of the standardized contrast scores are the sums of squares for the linear × subjects, quadratic × subjects, and cubic × subjects error terms of our analysis of variance. This analysis also shows that the F's for the linear, quadratic, and cubic trends are all equal to 1.00. We now want to show that we could directly compute a t test to test each of these contrasts if we preferred.

We begin by returning to the original data for three subjects (Table 5.9), since to compute t we shall want to operate on a set of L scores based on the original data (with or without subtracting off the grand mean, here equal to zero, and with or without subtracting off the row effects, here also equal to zero). For these data, the L scores for linear trend for the three subjects are 0, 20, and 0. Each L score is the sum of the products of the subject's scores multiplied by the contrast weight or λ, for example, $(-1)(-3) + (3)(-1) + (-3)(+1) + (1)(+3) = 0$. For these three L scores, the mean or $\bar{L} = 6.67$, $S^2 = 133.33$, $N = 3$, so that

$$t = \frac{\bar{L} - 0}{\sqrt{S^2/N}} = \frac{6.67}{\sqrt{133.33/3}} = 1.00$$

Since $t^2 = F$, we see that this method of computing contrasts yields results identical to those of the method of the analysis of variance shown in Table 5.10. Note, however, that the mean square for t, or S^2, is 133.33, which is $\Sigma\lambda^2$ or 20 times larger than the mean square error for the F of our table of variance (6.67).

For these data, the L scores for quadratic trend for the three subjects are 0, 0, and 4. Each score is the sum of the products of the subject's scores multiplied by the contrast weight or λ, for example, $(-1)(-1) + (3)(+1) + (-3)(+1) + (1)(-1) = 0$, $\bar{L} = 1.33$, $S^2 = 5.33$, $N = 3$, so that

$$t = \frac{1.33 - 0}{\sqrt{5.33/3}} = 1.00$$

Once again $t = 1 = t^2 = F$ showing that the method of t yields results identical to those based on the analysis of variance.

The L scores for the cubic trend for these three subjects are 20, 0, and 0. Once again, these L scores are computed as a sum of products, e.g., $(-1)(-1) + (3)(+3) + (-3)(-3) + (1)(+1) = 20$, $\bar{L} = 6.67$, $S^2 = 133.33$, $N = 3$, so that

$$t = \frac{6.67 - 0}{\sqrt{133.33/3}} = 1.00$$

Again, the method of t yields results identical to those of the analysis of variance procedure.

Effect size estimation

Whenever reporting the results of a contrast, some indication of effect size should also be given. When the contrasts are trend effects perhaps the most natural effect size indicator is r. We can compute r in several ways:

$$r = \sqrt{\frac{SS \text{ trend}}{(SS \text{ trend} + SS \text{ error for trend})}}$$

$$r = \sqrt{\frac{F}{F + df \text{ error for trend}}}$$

$$r = \sqrt{\frac{t^2}{t^2 + df}}$$

For all three of the t tests above, and for all the F's associated with each of these t's, $r = .58$. The *df* associated with each t, or with the denominator of each F, was 2. Suppose, however, that we had decided to use as our error terms in the analysis of variance not the unique error term but a pooled error term, on the assumption that the three error terms of 6.67, 1.33, and 6.67 differed from one another only by virtue of sampling variation. The pooled error term would then be $(13.33 + 2.67 + 13.33)/(2 + 2 + 2) = 4.89$. Our F's would now change from $F = 1.00$ for all three trend contrasts to F linear $= 1.36$; F quadratic $= 0.27$; and F cubic $= 1.36$. When looking up the significance level of these F's, we would enter Table A.4 in the Appendix, with 1 *df* for the numerator (all contrasts have $df = 1$ for the numerator of F) and with 6 *df* for the denominator, because we now have a more stable estimated error term, that is, one that is several degrees of freedom closer to the population value of the error term. However, and here it is easy to go astray, we should *not* use the pooled *df* in the formulas for computing r from F or t, but rather the *df* of the specifically computed error terms (e.g., linear × subjects). The formulas for computing r from F or t require that the *df* reflect the size of the experiment (i.e., the number of subjects minus the number of different between-subjects conditions) not the *df* for the estimate of the *MS* error. Often, of course, these two values will be the same. However, in repeated-measures studies having three or more levels of the repeated-measures variable, the *df*'s for various error terms (repeated measures × subjects interactions) will be larger than the *df* for specific error terms (e.g., linear trend × subjects interaction) by a factor equal to the number of levels of the repeated-measures factor minus one. The use of the *df* of the pooled error term can lead to estimates of effect size (e.g., r) that are much too low.

For our example of three subjects observed on four occasions, the three F's for contrasts (each computed on the basis of its specific error term) were all equal to 1.00, and for each, $r = .58$. When we employed a pooled error term, our three F's became 1.36, 0.27, and 1.36 each, with 1 and 6 *df* for the purpose of looking up the p value associated with each F. Using those *df* to compute r (an improper procedure) would give r's of .43, .21, and .43, respectively. However, using the proper *df* of 2 (rather than 6) for the denominator of F, or for t, would give r's of .64, .34, and .64.

Similarly, when using sums of squares to compute *r*, that is,

$$r = \sqrt{SS \text{ trend}/(SS \text{ trend} + SS \text{ error for trend})}$$

the *SS* for error for trend should either be the specific error term for that trend (e.g., linear × subjects interaction) or, if based on a pooled error term, it should be redefined as *MS* pooled error × *df* for specific error term, so that

$$r = \sqrt{SS \text{ trend}/(SS \text{ trend} + [df \text{ specific error term}]\ MS \text{ pooled error term})}.$$

Before leaving the topic of effect size estimation, we add a note of clarification as to why *df* for pooled error terms are appropriate to employ for purposes of looking up *p* values associated with test statistics but not for computing effect sizes. Imagine an experiment in which we compare a new with an old treatment procedure but we have available only four subjects. We assign two to each treatment procedure at random and find $t(2) = 2$, $F(1,2) = 4$, and $r = .82$, a very large effect size. Now imagine that we actually knew the *population* variance of the dependent variable we had employed. The *value* of *t* (or *F*) would not be affected, but now we would use a different row in our table of *t* (or *F*) to look up our significance levels, since now we have a *t* with $df = \infty$ rather than $df = 2$.

The *size of the effect,* however, should not be affected by whatever part of the table we employ to find the *p* value associated with *t* (or *F*). Yet, if we employed as our *df* for *t* (or for the error term of *F*) a very large number (like ∞), the actual *r* of .82 would be reduced to zero, an obviously absurd result. Table 5.13 shows, for increasing *df* for the error term, the effect on *r* if we were erroneously to employ pooled error *df* in the computation of *r* rather than the correct value of *df* reflecting the actual size of the experiment. In this example, that *df* was 2.

Homogeneity of covariances

A very important bonus accrues to us from our systematic use of contrasts in the case of repeated-measures analyses. In such analyses it is assumed that the various levels of the repeated measurements are all related to one another to a similar degree.

Table 5.13. *Effects of employing incorrect* df *for error term on computations of* r

df for Error Term	*F* or t^2	Correct r (df = 2)	Incorrect r
2	4	.82	–
4	4	.82	.71
8	4	.82	.58
16	4	.82	.45
32	4	.82	.33
64	4	.82	.24
128	4	.82	.17
256	4	.82	.12
512	4	.82	.09
1024	4	.82	.06
2048	4	.82	.04
4096	4	.82	.03
8192	4	.82	.02
∞	4	.82	.00

For example, if we measure a group of subjects 10 times on some task, the ordinary use of the omnibus *F* test with 9 *df* in the numerator requires that the 10th measurement correlates about as highly with the first as with the 9th, an assumption that may often not be met (Winer, 1971).

The bonus from the use of contrasts is that contrast effects, when tested against their specific contrast × subjects error terms (or against a pooled error term of relatively homogeneous components), do not require meeting such an assumption. Indeed, as we have seen, such contrasts are identical to *t* tests on subjects' contrast scores and there can be no covariance for just a single set of scores.

6
Mixed sources of variance

Between and within

We have had a detailed look at carving contrasts out of repeated measures. It is frequently the case, however, that repeated-measures designs occur in a context of mixed between-subjects and within-subjects factors. To anticipate the discussion a bit, this poses no new problem so long as the contrasts of interest are entirely part of a between-subjects factor or entirely part of a single within-subjects factor. It does pose a problem when the contrast of interest involves a between *and* a within factor, or two or more within-subjects factors having different error terms.

Computational example

For our illustration we return to the data of Table 5.1 of the last chapter, but with the added information that subjects A and B were female while subjects C and D were male. This additional information is contained in Table 6.1. The full model analysis of

Table 6.1. *Two groups of subjects each measured on three occasions*

Subjects		Occasions 1	2	3	Σ
Females	A	8	12	10	30
	B	7	11	9	27
Males	C	3	5	1	9
	D	2	4	0	6
Σ		20	32	20	72

Table 6.2. *Analysis of variance of data of Table 6.1*

Source	*SS*	*df*	*MS*	*F*	*p*
Between subjects	(150)	(3)	(50.00)		
Gender	147	1	147.00	98.00	.01
[a]Subjects (within gender)	3	2	1.50		
Within subjects	(32)	(8)			
Occasions	24	2	12.00	∞	1/∞
Occasions × gender	8	2	4.00	∞	1/∞
[b]Occasions × subjects	0	4	0		
Linear trend	0	1	0	—	—
Linear × gender	8	1	8.00	∞	1/∞
[b]Linear × subjects	0	2	0		
Quadratic trend	24	1	24.00	∞	1/∞
Quadratic × gender	0	1	0	—	—
[b]Quadratic × subjects	0	2	0		

[a]Appropriate error term for the just preceding effect.
[b]Appropriate error term for the just preceding two effects.

variance of these data is shown in Table 6.2. This table is analogous to that of Table 5.8 in the preceding chapter, except that we have added several sources of variance associated with our new factor of gender: gender, occasions × gender, linear trend × gender, and quadratic trend × gender. The latter two terms are contrasts carved out of the occasions × gender source of variance. There is nothing new about the contrast for gender; it was computed with weights (λ's) of $+1$ for females and -1 for males and tested against subjects within gender. The linear and quadratic contrasts were taken from the occasions effect, and the linear × gender and quadratic × gender contrasts were taken from the occasions × gender source of variance. All contrasts were tested against their own specific error terms. The weights for the linear × gender contrast were:

$$\begin{matrix} -1 & 0 & +1 \\ +1 & 0 & -1 \end{matrix}$$

Table 6.3. *Residuals for gender × occasions data*

	Occasions			
	1	2	3	Σ
Females	$-2^{\underline{\mid 2}}$	0	+2	$0^{\underline{\mid 6}}$
Males	+2	0	−2	0
Σ	$0^{\underline{\mid 4}}$	0	0	$0^{\underline{\mid 12}}$

Table 6.3 shows the residuals to which the weights just above were applied, so that $L^2/n\Sigma\lambda^2 = 8^2/2(4) = 8$; since $L = -2(-1) + 0(0) + 2(+1) + 2(+1) + 0(0) + -2(-1) = 8$. The weights for the quadratic × gender contrast were:

$$\begin{array}{ccc} -1 & +2 & -1 \\ +1 & -2 & +1 \end{array}$$

Complicating the design

So far there is nothing new or complicated about employing contrasts in mixed source designs. That is because we have only considered contrasts that had straightforward error terms. But suppose it were hypothesized that women tested on the first occasion would score higher than men tested on the third occasion. For this hypothesis, our weights (λ's) are shown in Table 6.4.

Table 6.4. *Weights (λ's) for the hypothesis that women tested first will score higher than men tested third*

	Occasions				
	1	2	3	Σ	Mean
Females	+1	0	0	+1	+ .33
Males	0	0	−1	−1	− .33
Σ	+1	0	−1	0	.00
Mean	+.50	.00	− .50	.00	.00

Table 6.5. *Data in Table 6.4 with row means removed*

	Occasions 1	2	3	Σ[a]	Mean
Females	.67	−.33	−.33	0	0
Males	.33	.33	−.67	0	0
Σ	1.00	.00	−1.00	0	0
Mean	.50	.00	−.50	0	0

[a] Rounded

Examination of the column totals or means shows that our contrast involves a prediction of a linear trend in occasions. The row totals show that our contrast also involves a prediction of a gender effect. The residuals show that our contrast also involves a prediction of an occasions × gender interaction, specifically a quadratic trend × gender interaction. We can show the quadratic trend × gender interaction that is implied by our contrast weights given in Table 6.4 simply by subtracting the row and column effects implied by our contrast. Thus, when subtracting the row means from these data, we have Table 6.5.

Then, subtracting the column means yields the residuals shown in Table 6.6. Inspection of the residuals in this table reveals that we are predicting a relative U shape for females and an inverted

Table 6.6. *Data in Table 6.5 with column means removed*

	Occasions 1	2	3	Σ[a]	Mean
Females	.17	−.33	.17	0	0
Males	−.17	.33	−.17	0	0
Σ	0	0	0	0	0
Mean	0	0	0	0	0

[a] Rounded

U shape (or ∩) for males. This quadratic trend × gender interaction, like all interaction effects, is unrelated to the row and column effects.

We *could* test for significance all the effects we had implicated in our contrast, but that would be a very low power procedure under most circumstances (essentially dissipating the strength of our prediction over three other contrasts). It would be better to test our hypothesis with just a single contrast. Rewriting Table 6.1 gives us the new totals shown in Table 6.7.

The contrast sum of squares, $L^2/n\Sigma\lambda^2$, is, therefore, $[15(+1) + 0 + 0 + 0 + 0 + 1(-1) = 14]^2/2(2) = 49$. Now we have the numerator for our *F*, but where shall we find the denominator? We have seen from our analysis of the column effects, row effects, and interaction effects that all are involved in our prediction so that the error terms for these three effects are *all* candidates for our choice of error term (i.e., subjects within gender; linear × subjects; and quadratic × subjects). The general approach to this problem involves four steps:

1 Diagnosis of effects involved
   A. Select all effects of the analysis due to the contrast
   B. Select all error terms for these effects
2 Aggregation where possible
   A. Total aggregation
   B. Partial aggregation
3 Consensus of high and low power tests
   A. Both yield significant results
   B. Neither yield significant results
4 Reporting of inconsistent tests

Table 6.7. *Rewritten table of totals based on the data in Table 6.1*

	Occasions			
	1	2	3	Σ
Females	15 ⌊2	23	19	57 ⌊6
Males	5	9	1	15
Σ	20 ⌊4	32	20	72 ⌊12

1. Diagnosis of effects

Once we have decided on our contrast, we diagnose any effects our contrast may have on any of the terms of our analysis model. In our example, we saw that our contrast had effects on one between-subjects main effect, one within-subjects main effect, and one within-subjects interaction effect. If we cannot be sure of these effects by inspection, we compute an analysis of variance directly on our contrast weights (λ's) and note any non-zero effects.

After we have diagnosed these non-zero effects we find the appropriate error term for each of them. It is these error terms we shall consider for aggregation.

2. Aggregation procedures

The basic idea in aggregation is to decide when to combine two or more error terms because (a) no prior theory suggests strongly that they should be different in their population values and (b) our estimates appear not too different, say within a factor of two of each other (Green and Tukey, 1960).

Suppose, for example, that in a study of the type we have been discussing we find the following *SS*'s, *df*, and *MS*'s for our three error terms.

Source	*SS*	*df*	*MS*		
Subjects within gender	200	2	100		aggregated = $\frac{200 + 160 + 120}{2 + 2 + 2} = 80$
Linear × subjects	160	2	80	aggregated = $\frac{160 + 120}{2 + 2} = 70$	
Quadratic × subjects	120	2	60		

We would begin at the bottom and work our way up, first trying to aggregate the repeated-measures error terms of 80 and 60.[1] Since $80/60 = 1.33$, far less than 2.0, we aggregate these terms by adding their sums of squares and dividing this grand sum by

[1] We begin at the bottom assuming that we have listed our sources of variance such that more complex error terms will be listed below simpler error terms. This is the usual format for listing sources of variance and is standard for most computer packages.

the sum of the *df*'s of the two terms. With equal *df* we arrive at a pooled error term for within-subjects of 70. Next we consider for aggregation the between-subjects term of $MS = 100$ with our newly aggregated $MS = 70$. This, too, seems aggregable since $100/70 = 1.43$, considerably less than two. Again we add the sums of squares of the candidates for aggregation and divide by the sum of their *df*'s. This was an example of total aggregation where all relevant error terms could be aggregated. The data of the analysis of variance table of this chapter are probably aggregable even though the *MS* for between-subjects error differs from the *MS* for within-error by a factor greater than 2.0. That is because the *df* for each error term are so small that we would greatly doubt that an error term could really be zero, and because a ratio of a larger to a smaller *MS* could reach 10, 20, or 30 before we would have very strong reason to believe the *MS*'s to differ in the population.

Partial aggregation occurs when some of the terms, usually some of the repeated-measures (within-subjects) mean squares, can be aggregated with one another but others (often a between-subjects term) cannot be. In that case we go on to step 3 in just the same way we would if no aggregation had been possible.

3. Consensus tests

Where total aggregation is possible, we have an easy answer to our question of which error term to employ for our contrast mean square. However, where no aggregation is possible, or where aggregation is only partial, no easy answer is available and we turn to consensus tests.

The basic idea of consensus tests is to employ as the denominator of our contrast mean square first the largest of the error terms involved in the contrast, and then the smallest. Employing the largest of the error terms yields an F for our contrast that we know to be too small; employing the smallest of the error terms yields an F that we know to be too large. However, if both F's lead to the conclusion that there probably is a relationship in the population between our independent and dependent variable (for example, if both F's are significant at our favorite alpha level), there is little problem in our having been unable to specify clearly a single error term.

Similarly, when the use of the largest and the smallest relevant error terms both lead to the same conclusion – that there is probably no appreciable relationship between our independent and dependent variable (for example, if both F's are clearly smaller than required for our favorite alpha level) – there is also little problem in our having been unable to specify clearly a single-error term.

4. *Inconsistent tests*

If the use of both the largest and smallest relevant error terms lead us to the same conclusion, there is no great problem from our having two or more candidates for a relevant error term for our contrast. However, it may happen instead that the use of our largest error term leads to a nonsignificant F while the use of our smallest error term leads to a significant F. In that case we shall have to give both F's along with the report that our results are ambiguous with respect to significance testing. If we find it essential to come up with a rough estimate of a single F, we can aggregate all the relevant error terms despite their heterogeneity and compute the F based on this new error term. However, we would employ the *df* for the denominator of our F test associated with the error term having the fewest *df* (Rosenthal & Rubin, 1980). This F may be only a poor estimate of the one we would like to be able to compute. Perhaps we can take some comfort from knowing that this estimate will at least be better than the two extreme F's we computed first, those based on the largest and smallest error terms.

Effect size estimation

Whichever of the above procedures we have followed to obtain the F test for our contrast, some estimate of the effect size should be reported for every significance test reported. The same procedures can be used that were described earlier but with the reminder that the *df* used in computing an effect size, say r, should be the *df* of the contrast-specific error term, *not* the *df* of the pooled error term. The appropriate *df*, then, will generally be the number of subjects minus the number of different between-subject conditions.

7
Practical issues of computations

Secondary analysis of data

So far in our discussion of contrasts we have assumed that the data collected were our own and that we had available for computation all the ingredients required to obtain the contrast sum of squares, which include the condition totals (T) required to compute L^2 and the n upon which each T is based. Then, after computing the sum of squares (or equivalently, the mean square) for the contrast by means of $L^2/n\Sigma\lambda^2$, it was necessary that we have available the mean square for error against which to test the contrast mean square.

Increasingly, however, behavioral and social researchers are engaging in secondary analyses of other people's data, sometimes based not on access to the original data but only on access to published reports (Rosenthal, 1980, 1984). In those instances the raw ingredients required to compute contrasts may not be reported. Indeed, it is a rare research report nowadays that provides the mean square for error employed for computing the various F ratios. Normally, it is possible to count on the provision of both the F ratios and the df with which each F is associated. Such information is required, for example, by journals following the *Publication Manual* of the *American Psychological Association* (APA, 1983). Whenever F's are reported in this manner, it is possible to compute contrasts among the means whose variation has been summarized by the reported F.

The logic of the procedure is simple. One computes a "maximum possible contrast F" (MPC-F) by multiplying the relevant F by its numerator df. This quantity represents the largest value

Table 7.1. *Results of testing four age groups on their ability to decode nonverbal cues*

Mean sensitivity scores for four age levels

Ages			
8	10	12	14
2.0	5.0	7.0	8.0

Analysis of variance of sensitivity scores

Source	*SS*	*df*	*MS*	*F*	*p*
Groups	210	3	70	2.0	>.10
Subjects within groups	1260	36	35		

of *F* that any contrast carved out of the sum of squares for the numerator of *F* could achieve, and it could achieve this value only if *all* of the variation among the means tested by the overall *F* were associated with the contrast computed. The *F* for our contrast is simply computed by multiplying the MPC-*F* by the proportion of variance among the means in question that is accounted for by the planned contrast. Suppose, for example, that a developmental psychologist had tested four age groups on their ability to decode nonverbal cues and had obtained the results shown in Table 7.1.

The maximum possible contrast *F* is $2.0 \times 3 = 6.0$. The proportion of variance among the means accounted for by the linear contrast is

$$\frac{L^2/n\Sigma\lambda^2}{SS\text{ means}} = \frac{[2(-3) + 5(-1) + 7(+1) + 8(+3)]^2/(1)\,(20)}{[2^2 + 5^2 + 7^2 + 8^2] - (22)^2/4}$$

$$= \frac{20}{21} = .952$$

Note that in these computations we have worked only with the four means shown above. We are not working with the sums of the four conditions, which we worked with earlier when we were computing contrasts in our own data. When our computations are done on the basis of others' reports of their data, the quantity n of $n\Sigma\lambda^2$ is equal to 1.

Depending on the type of calculator we have available, we may find it more convenient to compute $[L^2/n\Sigma\lambda^2]/[SS \text{ means}]$ as the square of the correlation coefficient r between our contrast weights (λ's) and the reported means. In this example, $r = .976$ and $r^2 = .952$. The latter is the same value we obtained by computing the contrast sum of squares based on just the four scores (means) shown in Table 7.1, divided by the sum of squares among those same four scores (means).

Whichever method of computing the proportion of variance we employ, $[L^2/n\Sigma\lambda^2]/[SS \text{ means}]$ or r^2, we need only multiply this value by our MPC-F to obtain our F for the contrast of interest.

For this example, F for our contrast was $.952 \times 6.0 = 5.71$, which for 1 and 36 df is significant at about the .02 level. This procedure of computation based on means of conditions is especially useful since totals per condition are often not estimable from others' reports of their research. It should be noted, however, that the same F values would result if we did know the totals for each cell and applied the computational procedures we originally introduced in our discussion of contrasts.

For the present data let us assume an equal n of 10 per cell, then

$$L^2/n\Sigma\lambda^2 = [20(-3) + 50(-1) + 70(+1) + 80(+3)]^2/(10)20$$
$$= (200)^2/200 = 200.$$

We divide this sum of squares (also a mean square since $df = 1$) by the mean square for error, which in this example was 35; $200/35 = 5.71$. This is the same value obtained just above by the method of taking a proportion of the maximum possible contrast F. We should also note that the sum of squares of 200 for the contrast, when divided by the sum of squares for the total group effect of 210, yields $200/210 = .952$. We recognize this as the value of the r^2 calculated between the mean of the cells and the contrast weights.

The procedures for calculating contrasts from other people's data (e.g., $r^2 \times$ MPC-F = contrast F) work well for any single source of variance from which the contrast is to be carved. Any single main effect or any single interaction can be analyzed for contrasts in this way. Working with main effects we can operate directly on cell means. Or we could operate equally well on the residuals, that is, the cell means from which the grand mean has been subtracted. Working with interactions, however, we operate *only* on the residuals defining interaction, that is, the cell means from which have been subtracted the grand mean, the row effects, and the column effects in the case of a two-way interaction. For higher-order interactions, all the main effects and interactions entering into the higher-order interaction must also be subtracted from the cell means in order to carve a contrast specifically out of the higher-order interaction without risking taking some of the contrast out of other and unwanted sources of variance.

If we want to take a contrast out of a two-way interaction, we can easily subtract off the grand mean, row effects, and column effects to obtain the needed residuals upon which to operate. With higher-order interactions, computing the residuals becomes more complex. We are also less likely to be given all the data we need to compute these residuals in the published reports of other workers. When contrasts in higher-order interactions are required, we should consider the use of data analytic packages that provide the residuals as a normal part of their output, for example, Data-Text developed by Armor and Couch (1972).

We can go further and say that computer statistical packages that do not provide for displays of residuals make it very difficult for investigators ever to see or interpret the higher-order interactions that may have been predicted or that turn out to be highly significant. As alluded to in Chapter 1, the common practice of comparing *means* to investigate the action of an interaction is ordinarily inappropriate, since these means are not due to single interaction effects alone but to many other effects as well.

Complex computer outputs

Sometimes the data are our own but considerations of computational convenience suggest that it is wisest to approach contrasts in the manner just described. With small data sets, equal sample sizes, and simple experimental designs, any computational method will work. Experience with our own data and experience in consulting over a period of years, however, suggest that with large data sets, unequal sample sizes, and complex experimental designs, the proportion of variance $\times$ MPC-F approach leads to fewer errors than does the basic definitional method of computing the contrast effect as $L^2/n\Sigma\lambda^2$ and dividing this quantity by the mean square for error to obtain the F for contrast.

The source of the errors is most often the calculation of the n in the formula for $L^2/n\Sigma\lambda^2$. In complex designs we are most likely to be working from computer output that has provided us with the means, hopefully the residuals defining the various interactions, and the table of variance. The n per condition in these cases is based upon the harmonic mean of the number of sampling units in each cell of the between-subjects design and the total number of times each unit has been measured in the repeated-measures dimensions of the design. The required n is computed by multiplying the harmonic mean of the number of sampling units in each cell of the between-subjects design by the number of levels in each between-subjects factor and the number of levels in each within-subjects factor, all divided by the number of weights employed for the contrast.

For example, suppose we have a $3 \times 4 \times 5 \times 3 \times 4$ design in which the first three factors were between-subjects factors and the last two were within-subjects factors. There would then be $3 \times 4 \times 5 = 60$ between-subjects cells and $3 \times 4 = 12$ measurements per subject. If the harmonic mean of the number of subjects in each cell were 2.5, we would have the equivalent of a total of $2.5 \times 60 \times 12 = 1800$ observations in the study. If we planned a contrast among the three levels of the between-subjects factor, n for each total would be $1800/3 = 600$. If we planned a contrast among the residuals defining a 4×5 interaction, 20 weights would be involved and n would be $1800/20 = 90$. If we planned a contrast among the residuals of a $4 \times 5 \times 4$ interaction, 80 weights would be involved and n would be $1800/80 = 22.5$. Getting the n right in a complex design is not a trivial task, and that is a major reason for our recommending the proportion of variance $\times$ MPC-F approach instead.

The quantity, n, is needed not only in the denominator of $L^2/n\Sigma\lambda^2$ but also in the numerator. Since L is the sum of the products of each *total* multiplied by its weight (λ), and since our computer printout gives us only means or residuals from means, each of these means or residuals must be multiplied by n to yield the total. If this method of computing contrasts is used, it would be wise to check the resulting F by the method of proportion of variance × MPC-*F*. The only advantage of employing the direct computational formula based on totals rather than the proportion of variance methods based on means is that the sum of squares computing from totals will be on the same scale as the sums of squares of the computer printout, since programs of analysis of variance (such as Data-Text) generally operate on totals rather than means.

Multiple sources of variance

So far in our discussion of practical computation procedures we have dealt only with contrasts carved from a single source of variance; but these procedures can be applied equally well to contrasts taken from two or more sources. Suppose, for example, that on the basis of our own research, or from reports of others' research, we were interested in the six means shown in Table 7.2.

Suppose further that we knew that the F's for treatment, age, and treatment-by-age interaction were 2.06, 0.05, and 0.39 respectively. We could still apply the proportion of variance × maximum possible contrast F approach to compute the F for any desired contrast. Suppose our theory called for the following contrast among the six means in Table 7.2:

	Low	Medium	High
Young	−1	0	+1
Old	0	0	0

that is, our theory predicts that the high level of treatment will be superior to the low level of treatment for young patients. To compute our F for the contrast, we will now need a maximum

Table 7.2. *Mean adjustment scores for three levels of treatment*

	Treatment levels			
Age	Low	Medium	High	Σ
Young	3	5	8	16
Old	4	5	6	15
Σ	7	10	14	31

possible contrast F based not on a single source of variance but on three sources of variance. As long as all three sources of variance can be tested against the same error term (e.g., as in a completely between-subjects design), we can compute an aggregated MPC-F simply by adding up the MPC-F's for all of the terms contributing to the variation among the cell means of interest to us. In the present example we have three terms contributing to the variation among the six means: treatments, ages, and treatment × ages interaction. We compute the MPC-F for each of these terms by multiplying each F by its numerator *df*. Our three MPC-F's, then, are 2.06 × (2 *df*), 0.05 × (1 *df*), and 0.39 × (2 *df*). When added together, they yield an MPC-F aggregated of 4.95. To obtain the F for our contrast, we will multiply this aggregate MPC-F by the proportion of variance among the means that is accounted for by the planned contrast.

For example, we could multiply 4.95 by the r^2 between the means and the contrast weights. In this example that $r^2 = .843$ and so the F for our contrast $= .843 \times 4.95 = 4.17$ ($p = .05$, if the *df* for the denominator of the F were at least 30). We will usually know the *df* for the error term if authors have followed APA style, but we may not know the n per condition of the experiment – because n may vary from cell to cell, and we may not know the harmonic mean n.

An alternative way to compute the proportion of variance required is by means of the formula

$$\frac{L^2/n\Sigma\lambda^2}{SS \text{ means}}$$

which, for this example, is

$$\frac{[3(-1) + 0 + 8(+1) + 0 + 0 + 0]^2/(1)(2)}{[3^2 + 5^2 + 8^2 + 4^2 + 5^2 + 6^2] - (31)^2/6} = \frac{12.50}{14.83} = .843$$

This value of .843 is equal to that obtained by computing r^2 but is easier to compute with calculators that only accumulate X, X^2, Y, Y^2, and XY without having a special key for computing r. Computing r^2 is easier than computing $[L^2/n\Sigma\lambda^2]/[SS \text{ means}]$ when our calculator does have a key for r.

Careful study of the contrast weights we chose in this example shows that part of this contrast involved a main effect of columns, and part involved the column × row interaction. However, the contrast did not involve any effect of rows. It might, therefore, be asked whether we could have managed by aggregating just those MPC-F's that contributed to effects that were predicted to be nonzero. Actually, we could have done so but it would have made our task a bit more complicated. The reasoning is as follows: When we want to take a contrast out of an interaction effect alone, we must remove the row and column effects from the means upon which we operate. Similarly, when we want to operate exclusively on any two of these three effects (row, column, and interaction), we must remove the effects of the third source of variance from the means upon which we operate.

For the example we have been considering, when we remove the row effect our table of means (with the grand mean left intact) becomes:

	Low	Medium	High	Σ
Young	2.83	4.83	7.83	15.5
Old	4.17	5.17	6.17	15.5
Σ	7	10	14	31

Our new MPC-F aggregated becomes $2.06 \times (2\ df) + 0.39 \times (2\ df) = 4.90$. Our new r^2 becomes .852, which when multiplied by 4.90 yields an F for our contrast of 4.17. This is the same value we obtained earlier when we had employed three F's in aggregating our MPC-F. The F for contrast remained unchanged, but both the r^2 and the MPC-F did change. The latter decreased because we were collecting fewer sources of variance. However, r^2 increased because our contrast was allocated a larger share of the smaller variance total to be allocated. Either method, then, yields the correct answer. However, because it is easy to forget to remove unwanted sources of variance from our means, it is safer when operating on multiple sources of variance to operate on means rather than on residuals representing two or more effects.

We described in detail the method of computing F's for contrasts that involves multiplying (a) the proportion of variance among our means (or residuals from means) that is accounted for by our contrast by (b) the maximum possible contrast F. This procedure of employing MPC-Fs is important because often another investigator's F is all the information we have. However, even when we do have more information, such as the sums of squares from which the contrast is to be taken and the mean square for error against which it is to be tested, it is convenient to employ the proportion of variance approach. Thus, we need only multiply the sum of squares from which the contrast sum of squares is to be drawn by r^2 or $[L^2/n\Sigma\lambda^2]/[SS \text{ means}]$ and divide by the mean square for error to obtain the F for our contrast. In its most convenient form: F contrast $= (r^2)(SS \text{ effect})/(MS \text{ error for that effect})$. The *SS* effect may be either a single source such as a main effect or an interaction, or it may be an aggregated source such as the *SS* rows + *SS* columns + *SS* interaction, from which our contrast *SS* is to be drawn.

Multiple error terms

Earlier, in our discussion of contrasts on mixed sources (Chapter 6), we addressed the question of computing contrasts in which several error terms were involved in testing their significance. If we have available only the F's for various effects from which contrasts are to be drawn, but if the error terms are aggregable, we can do fairly well by still applying the methods of the preceding section (e.g., $r^2 \times$ MPC-F aggregated).

However, if the error terms are not reasonably aggregable we may have to compute a series of contrasts, each one drawn from only a single source of variance that addressed our general question even though it is not as direct a contrast as we might like. Suppose, for example, that the six means we examined earlier were not the means of a 3 × 2 factorial design that was fully a between-subjects model, but rather a mixed design in which we had three levels of a between-subjects factor and two levels of a within-subjects factor:

	Groups			
	Low	Medium	High	Σ
Time 1	3	5	8	16
Time 2	4	5	6	15
Σ	7	10	14	31

Our contrast weights we recall as:

	Low	Medium	High	Σ
Time 1	−1	0	+1	0
Time 2	0	0	0	0
Σ	−1	0	+1	0

These weights imply no row effect, but both a column effect and an interaction effect. If we had insufficient information about the error terms for the between and within factors, or if these error terms differed greatly, we might feel it much too risky to aggregate error terms. Instead we might simply compute the linear trend in columns implied in our contrast weights. Since the F for columns was, say 2.06, the MPC-F for any contrast in columns would be $2.06 \times 2\ df = 4.12$. The proportion of variance index $r^2 = .993$, so F for overall linear contrast is 4.09, $p = .05$ if df for error is about 40 or more. Using our alternate procedure for computing proportion of variance accounted for

$$\frac{L^2/n\Sigma\lambda^2}{SS \text{ means}} = \frac{7^2/(1)(2)}{24.67} = .993,$$

that is, an alternative computation of r^2.

Part of our planned contrast also implies some interaction effect. We can see precisely what our prediction about the interaction term was by subtracting off the column effects from our table of contrast weights. This yields:

	Low	Medium	High	Σ
Time 1	−.5	0	+.5	0
Time 2	+.5	0	−.5	0
Σ	0	0	0	0

To obtain the contrast exclusively from the interaction effect we first compute the residuals that define the interaction by subtracting off the row and column effects. If we use an r^2 method it will not matter whether the grand mean is removed or not, since adding a constant does not affect a correlation coefficient. The table of residuals, grand mean removed, is:

	Low	Medium	High	Σ
Time 1	−.67	−.17	+.83	0
Time 2	+.67	+.17	−.83	0
Σ	0	0	0	0

The r^2 between these residuals and the contrast weights of the preceding table is .964 which, when multiplied by MPC-F for the interaction, would give the F for the contrast taken exclusively from the interaction.

This procedure of computing a separate contrast for each available F is not recommended in general, since it tends to dissipate the power of our contrast over two or more sources of variance. We recommend it as a last resort when the only information we have available is the set of F's provided in a research report. In general, it is preferable to compute the single contrast most directly addressing our research question and described in detail in Chapter 6. In that case, however, it will be necessary to have more information available, such as the sums of squares for the various effects and the error terms for these effects.

8

Measuring the benefits of contrasts

Indexing usefulness

Throughout our discussion of contrasts we have emphasized the benefits in terms of power of employing focused rather than diffuse (or omnibus) tests of hypotheses, for example, F tests with one rather than more than one df in the numerator. In concluding this book we want to introduce some measures of the benefits of having employed a contrast in any particular study. The purpose of these measures is to quantify the gains in explanatory power that have accrued to us by our having shifted from an omnibus F test to a focused F test or contrast. The ideas will be clearer after we have examined Table 8.1, which gives a table of variance in which the factor from which the contrast has been carved is shown along with the contrast and noncontrast sources of variation that together make up the total effect.

In this example the SS of the total effect is based on 5 df and, assuming any contrast is as good as any other, we would expect any randomly constituted contrast to absorb its "fair share" of the SS of the total effect. The fair share is defined as $1/df$, in this

Table 8.1. *One-way analysis of variance showing contrast and non-contrast components of the total effect*

Source	SS	df	MS	F	p
Total effect	100	5	20	4	.004
Contrast	60	1	60	12	.001
Non-contrast	40	4	10	2	.11
Error	300	60	5		

case 1/5 or .20. Any randomly chosen contrast should be associated with 20% of the *SS* if there are 5 *df*, 10% if there are 10 *df*, or 50% if there are only 2 *df*. We will index the success, utility, or benefit of a planned contrast by the proportion of variance it accounts for that exceeds the expected value of the average contrast (1/*df*). In computing proportion of variance, we continue to employ the r^2 between the means obtained and the contrast weights, directly computed, or alternatively computed as

$$\frac{L^2/n\Sigma\lambda^2}{SS \text{ means}}.$$

Under the hypothesis of a randomly chosen contrast, or a not especially clever contrast, we expect our r^2 to equal 1/*df*, that is, the "fair share" that every contrast should obtain. Our contrast has benefited us in proportion to the excess of our r^2 to this expected value of r^2 of 1/*df*.

The *maximum possible benefit* for any given number of *df* would occur if our $r^2 = 1.00$, and it is therefore defined as $1.00 - 1/df$ or, rewritten, as $(df - 1)/df$. For the data of Table 8.1, this value is $(5 - 1)/5 = .80$.

The *net contrast benefit* is defined as the r^2 corrected for the portion thereof we would expect if our contrast were only randomly selected, or $r^2 - (1/df)$. This can be rewritten as $(df\,r^2 - 1)/df$. For the data of Table 8.1, this value is $(60/100) - (1/5)$ or $[(5)(.60) - 1]/5 = .40$. We can think of this quantity as the amount of information in the contrast *df* after subtracting off the information in the unselected average *df*, or as the *net information per contrast df*.

Comparing benefit measures

Table 8.2 gives examples of maximum possible benefit scores (column 2) and net contrast benefit measures for three values of r^2 (columns 3, 4, 5) for selected values of *df* from 2 to 100. Column 2 shows that, for very low *df*, the maximum possible benefit is noticeably smaller than it is for even a few more *df*. Columns 3, 4, and 5 show that the net contrast benefit measure increases not only as r^2 increases but as the *df* for the effect from which the contrast has been taken increases. The upper limit for the net contrast benefit measure achieved, as *df* grow very large, is r^2.

Table 8.2. *Comparison of contrast benefit measures*

	Maximum possible benefit	Net contrast benefit $(df\,r^2 - 1)/df$ given r^2 to be:			Net contrast benefit/ maximum possible benefit given r^2 to be:		
df	$(df - 1)/df$	.20	.50	.80	.20	.50	.80
2	.50	*	.00	.30	*	.00	.60
3	.67	*	.17	.47	*	.25	.70
4	.75	*	.25	.55	*	.33	.73
5	.80	.00	.30	.60	.00	.38	.75
6	.83	.03	.33	.63	.04	.40	.76
7	.86	.06	.36	.66	.07	.42	.77
8	.88	.08	.38	.68	.09	.43	.77
9	.89	.09	.39	.69	.10	.44	.78
10	.90	.10	.40	.70	.11	.44	.78
15	.93	.13	.43	.73	.14	.46	.78
20	.95	.15	.45	.75	.16	.47	.79
25	.96	.16	.46	.76	.17	.48	.79
50	.98	.18	.48	.78	.18	.49	.80
100	.99	.19	.49	.79	.19	.49	.80

*Negative value

Because the net contrast benefit measure (NCBM) is forced to remain fairly small when *df* are few, it may be useful to divide the NCBM by the maximum possible benefit score (MPBS) to obtain a proportion of possible benefit score (POPBS). The last three columns of Table 8.2 show this score for r's of .20, .50, and .80, respectively. For at least moderate numbers of *df*, say over 10, this index differs little from the NCBM. For small numbers of *df*, however, the POPBS is substantially larger than the "uncorrected" net contrast benefit measure. A convenient computational formula for the POPBS is $(df\,r^2 - 1)/(df - 1)$. An alternative way to conceptualize this measure is as the difference between the r^2 (between the obtained means and the contrast weights) and the average r^2 per *df* not involved in the contrast. This latter average r^2 is computed as $(1 - r^2)/(df - 1)$ and when that quantity is subtracted from r^2 we obtain $(df\ r^2 - 1)/(df - 1)$, the same formula as for the POPBS. For the data of the one-way analysis of variance we have been considering this value is $[5(.60) - 1]/4 = .50$.

The formulas we have given for the various contrast benefit measures have employed only *df* and r^2, quantities more likely to be available to us in secondary analyses than sums of squares or mean squares. If we do have the latter available, we can compute the POPBS by

$$\frac{MS_{\text{Contrast}} - MS_{\text{Total effect}}}{SS_{\text{Total effect}} - MS_{\text{Total effect}}}$$

that for our data yields $(60 - 20)/(100 - 20) = .50$. An alternative computational formula is

$$\frac{(df)SS_{\text{Contrast}} - SS_{\text{Total effect}}}{(df - 1)SS_{\text{Total effect}}}$$

that for our data yields $[(5)(60) - 100]/(4)(100) = .50$.

If we should have sums of squares or mean squares available, and wanted to compute not POPBS but the "uncorrected" net contrast benefit measure, we could do so from the following:

$$\frac{MS_{\text{Contrast}} - MS_{\text{Total effect}}}{(df)MS_{\text{Total effect}}}$$

that for our data yields $(60 - 20)/(5)(20) = .40$. An alternative formula for the NCBM is

$$\frac{(df)SS_{\text{Contrast}} - SS_{\text{Total effect}}}{(df)SS_{\text{Total effect}}}$$

that for our data yields $[(5)(60) - 100]/(5)(100) = .40$.

In our discussion of measures of the benefits of contrasts we have made no mention of testing the significance of these benefits. Significance testing is not our primary concern in assessing the benefits of contrasts. Our primary concern, rather, is to indicate our gains in information per *df* by having planned our contrasts instead of simply accepting a kind of average contrast which is estimated by the mean square for the effect based on $df > 1$. In case some test of significance were required, one is readily available and we need only know the *df* for the effect from which the contrast is taken and the r^2 between the condition means and the contrast weights. Thus, the quantity $r^2\ (df - 1)/(1 - r^2)$ is distributed as F with 1 and $df - 1$ as the numerator and denominator *df*, respectively. This test is very conservative, however, because (a) the denominator *df*'s are usually far fewer than the *df* for the *MS* for error for the overall effect and (b) this F employs as an error term for the contrast *MS* the *MS* for the residual between-conditions variance. The latter often contains fairly substantial effects other than our planned contrast and is, therefore, too large an error term, leading to F's that are too small.

9

Conclusion: Abelson's perspective

In a classic work on research methods, Eugene Webb, Donald Campbell, Richard Schwartz, and Lee Sechrest (1966) concluded their book with a seven word chapter that beautifully captured the spirit of their work.

Our final chapter will be a bit longer but it speaks so wisely of our topic that we can think of no better way to conclude our book. The quotation this time is from an unpublished paper by Professor Robert Abelson, a paper written nearly a quarter of a century ago (1962). Had the wisdom in that paper been recognized by journal editors when it first became available, our field's thinking about data analysis in general, and about analysis of variance in particular, would be further advanced than it is today.

Here is what Professor Abelson said about the "method of contrasts":

> This method dates back virtually to the invention of the analysis of variance itself. . . . It is well-known to most statisticians and to some psychologists, but it has received only the most cursory and off-hand treatment in standard statistical reference works . . . and presentations by psychologists of some of its uses have tended toward very specialized applications. . . . Actually, the method of contrasts is extraordinary for its wide range of varied uses. That the method has not heretofore received a comprehensive, unified treatment is a matter of some mystery. One compelling line of explanation is that the statisticians do not regard the idea as mathematically very interesting (it is based on quite elementary statistical concepts) and that quantitative psychologists have never quite appreciated its generality of application.

It is our hope that the present volume helps to foster that appreciation of the value of contrast analysis that Professor Abelson found missing nearly a quarter of a century ago.

Appendix: Statistical tables

Table A.1. *Weights for orthogonal polynomial-based contrasts*

		Ordered conditions									
*k**	Polynomial†	1	2	3	4	5	6	7	8	9	10
2	Linear	−1	+1								
3	Linear	−1	0	+1							
	Quadratic	+1	−2	+1							
4	Linear	−3	−1	+1	+3						
	Quadratic	+1	−1	−1	+1						
	Cubic	−1	+3	−3	+1						
5	Linear	−2	−1	0	+1	+2					
	Quadratic	+2	−1	−2	−1	+2					
	Cubic	−1	+2	0	−2	+1					
6	Linear	−5	−3	−1	+1	+3	+5				
	Quadratic	+5	−1	−4	−4	−1	+5				
	Cubic	−5	+7	+4	−4	−7	+5				
7	Linear	−3	−2	−1	0	+1	+2	+3			
	Quadratic	+5	0	−3	−4	−3	0	+5			
	Cubic	−1	+1	+1	0	−1	−1	+1			
8	Linear	−7	−5	−3	−1	+1	+3	+5	+7		
	Quadratic	+7	+1	−3	−5	−5	−3	+1	+7		
	Cubic	−7	+5	+7	+3	−3	−7	−5	+7		
9	Linear	−4	−3	−2	−1	0	+1	+2	+3	+4	
	Quadratic	+28	+7	−8	−17	−20	−17	−8	+7	+28	
	Cubic	−14	+7	+13	+9	0	−9	−13	−7	+14	
10	Linear	−9	−7	−5	−3	−1	+1	+3	+5	+7	+9
	Quadratic	+6	+2	−1	−3	−4	−4	−3	−1	+2	+6
	Cubic	−42	+14	+35	+31	+12	−12	−31	−35	−14	+42

* Number of conditions.
† Shape of trend.

Source: Reproduced from R. Rosenthal and R. L. Rosnow, *Essentials of Behavioral Research: Methods and Data Analysis,* McGraw-Hill, New York, 1984, p. 352, with the permission of the publisher.

Table A.2. *Table of standard normal deviates*

	Second digit of Z									
Z	.00	.01	.02	.03	.04	.05	.06	.07	.08	.09
.0	.5000	.4960	.4920	.4880	.4840	.4801	.4761	.4721	.4681	.4641
.1	.4602	.4562	.4522	.4483	.4443	.4404	.4364	.4325	.4286	.4247
.2	.4207	.4168	.4129	.4090	.4052	.4013	.3974	.3936	.3897	.3859
.3	.3821	.3783	.3745	.3707	.3669	.3632	.3594	.3557	.3520	.3483
.4	.3446	.3409	.3372	.3336	.3300	.3264	.3228	.3192	.3156	.3121
.5	.3085	.3050	.3015	.2981	.2946	.2912	.2877	.2843	.2810	.2776
.6	.2743	.2709	.2676	.2643	.2611	.2578	.2546	.2514	.2483	.2451
.7	.2420	.2389	.2358	.2327	.2296	.2266	.2236	.2206	.2177	.2148
.8	.2119	.2090	.2061	.2033	.2005	.1977	.1949	.1922	.1894	.1867
.9	.1841	.1814	.1788	.1762	.1736	.1711	.1685	.1660	.1635	.1611
1.0	.1587	.1562	.1539	.1515	.1492	.1469	.1446	.1423	.1401	.1379
1.1	.1357	.1335	.1314	.1292	1271	.1251	.1230	.1210	.1190	.1170
1.2	.1151	.1131	.1112	.1093	.1075	.1056	.1038	.1020	.1003	.0985
1.3	.0968	.0951	.0934	.0918	.0901	.0885	.0869	.0853	.0838	.0823
1.4	.0808	.0793	.0778	.0764	.0749	.0735	.0721	.0708	.0694	.0681
1.5	.0668	.0655	.0643	.0630	.0618	.0606	.0594	.0582	.0571	.0559
1.6	.0548	.0537	.0526	.0516	.0505	.0495	.0485	.0475	.0465	.0455
1.7	.0446	.0436	.0427	.0418	.0409	.0401	.0392	.0384	.0375	.0367
1.8	.0359	.0351	.0344	.0336	.0329	.0322	.0314	.0307	.0301	.0294
1.9	.0287	.0281	.0274	.0268	.0262	.0256	.0250	.0244	.0239	.0233
2.0	.0228	.0222	.0217	.0212	.0207	.0202	.0197	.0192	.0188	.0183
2.1	.0179	.0174	.0170	.0166	.0162	.0158	.0154	.0150	.0146	.0143
2.2	.0139	.0136	.0132	.0129	.0125	.0122	.0119	.0116	.0113	.0110
2.3	.0107	.0104	.0102	.0099	.0096	.0094	.0091	.0089	.0087	.0084
2.4	.0082	.0080	.0078	.0075	.0073	.0071	.0069	.0068	.0066	.0064
2.5	.0062	.0060	.0059	.0057	.0055	.0054	.0052	.0051	.0049	.0048
2.6	.0047	.0045	.0044	.0043	.0041	.0040	.0039	.0038	.0037	.0036
2.7	.0035	.0034	.0033	.0032	.0031	.0030	.0029	.0028	.0027	.0026
2.8	.0026	.0025	.0024	.0023	.0023	.0022	.0021	.0021	.0020	.0019
2.9	.0019	.0018	.0018	.0017	.0016	.0016	.0015	.0015	.0014	.0014
3.0	.0013	.0013	.0013	.0012	.0012	.0011	.0011	.0011	.0010	.0010
3.1	.0010	.0009	.0009	.0009	.0008	.0008	.0008	.0008	.0007	.0007
3.2	.0007									
3.3	.0005									
3.4	.0003									
3.5	.00023									
3.6	.00016									
3.7	.00011									
3.8	.00007									
3.9	.00005									
4.0*	.00003									

Note: All p values are one-tailed in this table.
*Additional values of Z are found in the bottom row of Table A.3 since t values for $df = \infty$ are also Z values.
Source: Reproduced from S. Siegel, *Nonparametric Statistics*, McGraw-Hill, New York, 1956, p. 247, with the permission of the publisher.

Table A.3. *Extended table of* t

df \ p	.25	.10	.05	.025	.01	.005	.0025	.001
1	1.000	3.078	6.314	12.706	31.821	63.657	127.321	318.309
2	.816	1.886	2.920	4.303	6.965	9.925	14.089	22.327
3	.765	1.638	2.353	3.182	4.541	5.841	7.453	10.214
4	.741	1.533	2.132	2.776	3.747	4.604	5.598	7.173
5	.727	1.476	2.015	2.571	3.365	4.032	4.773	5.893
6	.718	1.440	1.943	2.447	3.143	3.707	4.317	5.208
7	.711	1.415	1.895	2.365	2.998	3.499	4.029	4.785
8	.706	1.397	1.860	2.306	2.896	3.355	3.833	4.501
9	.703	1.383	1.833	2.262	2.821	3.250	3.690	4.297
10	.700	1.372	1.812	2.228	2.764	3.169	3.581	4.144
11	.697	1.363	1.796	2.201	2.718	3.106	3.497	4.025
12	.695	1.356	1.782	2.179	2.681	3.055	3.428	3.930
13	.694	1.350	1.771	2.160	2.650	3.012	3.372	3.852
14	.692	1.345	1.761	2.145	2.624	2.977	3.326	3.787
15	.691	1.341	1.753	2.131	2.602	2.947	3.286	3.733
16	.690	1.337	1.746	2.120	2.583	2.921	3.252	3.686
17	.689	1.333	1.740	2.110	2.567	2.898	3.223	3.646
18	.688	1.330	1.734	2.101	2.552	2.878	3.197	3.610
19	.688	1.328	1.729	2.093	2.539	2.861	3.174	3.579
20	.687	1.325	1.725	2.086	2.528	2.845	3.153	3.552
21	.686	1.323	1.721	2.080	2.518	2.831	3.135	3.527
22	.686	1.321	1.717	2.074	2.508	2.819	3.119	3.505
23	.685	1.319	1.714	2.069	2.500	2.807	3.104	3.485
24	.685	1.318	1.711	2.064	2.492	2.797	3.090	3.467
25	.684	1.316	1.708	2.060	2.485	2.787	3.078	3.450
26	.684	1.315	1.706	2.056	2.479	2.779	3.067	3.435
27	.684	1.314	1.703	2.052	2.473	2.771	3.057	3.421
28	.683	1.313	1.701	2.048	2.467	2.763	3.047	3.408
29	.683	1.311	1.699	2.045	2.462	2.756	3.038	3.396
30	.683	1.310	1.697	2.042	2.457	2.750	3.030	3.385
35	.682	1.306	1.690	2.030	2.438	2.724	2.996	3.340
40	.681	1.303	1.684	2.021	2.423	2.704	2.971	3.307
45	.680	1.301	1.679	2.014	2.412	2.690	2.952	3.281
50	.679	1.299	1.676	2.009	2.403	2.678	2.937	3.261
55	.679	1.297	1.673	2.004	2.396	2.668	2.925	3.245
60	.679	1.296	1.671	2.000	2.390	2.660	2.915	3.232
70	.678	1.294	1.667	1.994	2.381	2.648	2.899	3.211
80	.678	1.292	1.664	1.990	2.374	2.639	2.887	3.195
90	.677	1.291	1.662	1.987	2.368	2.632	2.878	3.183
100	.677	1.290	1.660	1.984	2.364	2.626	2.871	3.174
200	.676	1.286	1.652	1.972	2.345	2.601	2.838	3.131
500	.675	1.283	1.648	1.965	2.334	2.586	2.820	3.107
1,000	.675	1.282	1.646	1.962	2.330	2.581	2.813	3.098
2,000	.675	1.282	1.645	1.961	2.328	2.578	2.810	3.094
10,000	.675	1.282	1.645	1.960	2.327	2.576	2.808	3.091
∞	.674	1.282	1.645	1.960	2.326	2.576	2.807	3.090

Note: All *p* values are one-tailed in this table.

Table A.3. (*Continued*)

df \ p	.0005	.00025	.0001	.00005	.000025	.00001
1	636.619	1,273.239	3,183.099	6,366.198	12,732.395	31,830.989
2	31.598	44.705	70.700	99.992	141.416	223.603
3	12.924	16.326	22.204	28.000	35.298	47.928
4	8.610	10.306	13.034	15.544	18.522	23.332
5	6.869	7.976	9.678	11.178	12.893	15.547
6	5.959	6.788	8.025	9.082	10.261	12.032
7	5.408	6.082	7.063	7.885	8.782	10.103
8	5.041	5.618	6.442	7.120	7.851	8.907
9	4.781	5.291	6.010	6.594	7.215	8.102
10	4.587	5.049	5.694	6.211	6.757	7.527
11	4.437	4.863	5.453	5.921	6.412	7.098
12	4.318	4.716	5.263	5.694	6.143	6.756
13	4.221	4.597	5.111	5.513	5.928	6.501
14	4.140	4.499	4.985	5.363	5.753	6.287
15	4.073	4.417	4.880	5.239	5.607	6.109
16	4.015	4.346	4.791	5.134	5.484	5.960
17	3.965	4.286	4.714	5.044	5.379	5.832
18	3.922	4.233	4.648	4.966	5.288	5.722
19	3.883	4.187	4.590	4.897	5.209	5.627
20	3.850	4.146	4.539	4.837	5.139	5.543
21	3.819	4.110	4.493	4.784	5.077	5.469
22	3.792	4.077	4.452	4.736	5.022	5.402
23	3.768	4.048	4.415	4.693	4.972	5.343
24	3.745	4.021	4.382	4.654	4.927	5.290
25	3.725	3.997	4.352	4.619	4.887	5.241
26	3.707	3.974	4.324	4.587	4.850	5.197
27	3.690	3.954	4.299	4.558	4.816	5.157
28	3.674	3.935	4.275	4.530	4.784	5.120
29	3.659	3.918	4.254	4.506	4.756	5.086
30	3.646	3.902	4.234	4.482	4.729	5.054
35	3.591	3.836	4.153	4.389	4.622	4.927
40	3.551	3.788	4.094	4.321	4.544	4.835
45	3.520	3.752	4.049	4.269	4.485	4.766
50	3.496	3.723	4.014	4.228	4.438	4.711
55	3.476	3.700	3.986	4.196	4.401	4.667
60	3.460	3.681	3.962	4.169	4.370	4.631
70	3.435	3.651	3.926	4.127	4.323	4.576
80	3.416	3.629	3.899	4.096	4.288	4.535
90	3.402	3.612	3.878	4.072	4.261	4.503
100	3.390	3.598	3.862	4.053	4.240	4.478
200	3.340	3.539	3.789	3.970	4.146	4.369
500	3.310	3.504	3.747	3.922	4.091	4.306
1,000	3.300	3.492	3.733	3.906	4.073	4.285
2,000	3.295	3.486	3.726	3.898	4.064	4.275
10,000	3.292	3.482	3.720	3.892	4.058	4.267
∞	3.291	3.481	3.719	3.891	4.056	4.265

Note: All p values are one-tailed in this table.

Table A.3. (*Continued*)

df \ *p*	.000005	.0000025	.000001	.0000005	.00000025	.0000001
1	63,661.977	127,323.954	318,309.886	636,619.772	1,273,239.545	3,183,098.862
2	316.225	447.212	707.106	999.999	1,414.213	2,236.068
3	60.397	76.104	103.299	130.155	163.989	222.572
4	27.771	33.047	41.578	49.459	58.829	73.986
5	17.807	20.591	24.771	28.477	32.734	39.340
6	13.555	15.260	17.830	20.047	22.532	26.286
7	11.215	12.437	14.241	15.764	17.447	19.932
8	9.782	10.731	12.110	13.257	14.504	16.320
9	8.827	9.605	10.720	11.637	12.623	14.041
10	8.150	8.812	9.752	10.516	11.328	12.492
11	7.648	8.227	9.043	9.702	10.397	11.381
12	7.261	7.780	8.504	9.085	9.695	10.551
13	6.955	7.427	8.082	8.604	9.149	9.909
14	6.706	7.142	7.743	8.218	8.713	9.400
15	6.502	6.907	7.465	7.903	8.358	8.986
16	6.330	6.711	7.233	7.642	8.064	8.645
17	6.184	6.545	7.037	7.421	7.817	8.358
18	6.059	6.402	6.869	7.232	7.605	8.115
19	5.949	6.278	6.723	7.069	7.423	7.905
20	5.854	6.170	6.597	6.927	7.265	7.723
21	5.769	6.074	6.485	6.802	7.126	7.564
22	5.694	5.989	6.386	6.692	7.003	7.423
23	5.627	5.913	6.297	6.593	6.893	7.298
24	5.566	5.845	6.218	6.504	6.795	7.185
25	5.511	5.783	6.146	6.424	6.706	7.085
26	5.461	5.726	6.081	6.352	6.626	6.993
27	5.415	5.675	6.021	6.286	6.553	6.910
28	5.373	5.628	5.967	6.225	6.486	6.835
29	5.335	5.585	5.917	6.170	6.426	6.765
30	5.299	5.545	5.871	6.119	6.369	6.701
35	5.156	5.385	5.687	5.915	6.143	6.447
40	5.053	5.269	5.554	5.768	5.983	6.266
45	4.975	5.182	5.454	5.659	5.862	6.130
50	4.914	5.115	5.377	5.573	5.769	6.025
55	4.865	5.060	5.315	5.505	5.694	5.942
60	4.825	5.015	5.264	5.449	5.633	5.873
70	4.763	4.946	5.185	5.363	5.539	5.768
80	4.717	4.896	5.128	5.300	5.470	5.691
90	4.682	4.857	5.084	5.252	5.417	5.633
100	4.654	4.826	5.049	5.214	5.376	5.587
200	4.533	4.692	4.897	5.048	5.196	5.387
500	4.463	4.615	4.810	4.953	5.094	5.273
1,000	4.440	4.590	4.781	4.922	5.060	5.236
2,000	4.428	4.578	4.767	4.907	5.043	5.218
10,000	4.419	4.567	4.756	4.895	5.029	5.203
∞	4.417	4.565	4.753	4.892	5.026	5.199

Note: All *p* values are one-tailed in this table.
Standard normal deviates (Z) corresponding to t can be estimated quite accurately from:

$$Z = \left[df \log_e \left(1 + \frac{t^2}{df}\right)\right]^{1/2} \left[1 - \frac{1}{2df}\right]^{1/2}$$

Source: Reproduced from E. T. Federighi. Extended tables of the percentage points of Student's *t*-distribution. *Journal of the American Statistical Association*, 1959, *54*, 683–688, with the permission of the publisher.

Table A.4. *Extended table of* F

df_2 \ df_1	p	1	2	3	4	5	6	8	12	24	∞
1	.001	405284	500000	540379	562500	576405	585937	598144	610667	623497	636619
	.005	16211	20000	21615	22500	23056	23437	23925	24426	24940	25465
	.01	4052	4999	5403	5625	5764	5859	5981	6106	6234	6366
	.025	647.79	799.50	864.16	899.58	921.85	937.11	956.66	976.71	997.25	1018.30
	.05	161.45	199.50	215.71	224.58	230.16	233.99	238.88	243.91	249.05	254.32
	.10	39.86	49.50	53.59	55.83	57.24	58.20	59.44	60.70	62.00	63.33
	.20	9.47	12.00	13.06	13.73	14.01	14.26	14.59	14.90	15.24	15.58
2	.001	998.5	999.0	999.2	999.2	999.3	999.3	999.4	999.4	999.5	999.5
	.005	198.50	199.00	199.17	199.25	199.30	199.33	199.37	199.42	199.46	199.51
	.01	98.49	99.00	99.17	99.25	99.30	99.33	99.36	99.42	99.46	99.50
	.025	38.51	39.00	39.17	39.25	39.30	39.33	39.37	39.42	39.46	39.50
	.05	18.51	19.00	19.16	19.25	19.30	19.33	19.37	19.41	19.45	19.50
	.10	8.53	9.00	9.16	9.24	9.29	9.33	9.37	9.41	9.45	9.49
	.20	3.56	4.00	4.16	4.24	4.28	4.32	4.36	4.40	4.44	4.48
3	.001	167.5	148.5	141.1	137.1	134.6	132.8	130.6	128.3	125.9	123.5
	.005	55.55	49.80	47.47	46.20	45.39	44.84	44.13	43.39	42.62	41.83
	.01	34.12	30.81	29.46	28.71	28.24	27.91	27.49	27.05	26.60	26.12
	.025	17.44	16.04	15.44	15.10	14.89	14.74	14.54	14.34	14.12	13.90
	.05	10.13	9.55	9.28	9.12	9.01	8.94	8.84	8.74	8.64	8.53
	.10	5.54	5.46	5.39	5.34	5.31	5.28	5.25	5.22	5.18	5.13
	.20	2.68	2.89	2.94	2.96	2.97	2.97	2.98	2.98	2.98	2.98
4	.001	74.14	61.25	56.18	53.44	51.71	50.53	49.00	47.41	45.77	44.05
	.005	31.33	26.28	24.26	23.16	22.46	21.98	21.35	20.71	20.03	19.33
	.01	21.20	18.00	16.69	15.98	15.52	15.21	14.80	14.37	13.93	13.46
	.025	12.22	10.65	9.98	9.60	9.36	9.20	8.98	8.75	8.51	8.26
	.05	7.71	6.94	6.59	6.39	6.26	6.16	6.04	5.91	5.77	5.63
	.10	4.54	4.32	4.19	4.11	4.05	4.01	3.95	3.90	3.83	3.76
	.20	2.35	2.47	2.48	2.48	2.48	2.47	2.47	2.46	2.44	2.43
5	.001	47.04	36.61	33.20	31.09	29.75	28.84	27.64	26.42	25.14	23.78
	.005	22.79	18.31	16.53	15.56	14.94	14.51	13.96	13.38	12.78	12.14
	.01	16.26	13.27	12.06	11.39	10.97	10.67	10.29	9.89	9.47	9.02
	.025	10.01	8.43	7.76	7.39	7.15	6.98	6.76	6.52	6.28	6.02
	.05	6.61	5.79	5.41	5.19	5.05	4.95	4.82	4.68	4.53	4.36
	.10	4.06	3.78	3.62	3.52	3.45	3.40	3.34	3.27	3.19	3.10
	.20	2.18	2.26	2.25	2.24	2.23	2.22	2.20	2.18	2.16	2.13
6	.001	35.51	27.00	23.70	21.90	20.81	20.03	19.03	17.99	16.89	15.75
	.005	18.64	14.54	12.92	12.03	11.46	11.07	10.57	10.03	9.47	8.88
	.01	13.74	10.92	9.78	9.15	8.75	8.47	8.10	7.72	7.31	6.88
	.025	8.81	7.26	6.60	6.23	5.99	5.82	5.60	5.37	5.12	4.85
	.05	5.99	5.14	4.76	4.53	4.39	4.28	4.15	4.00	3.84	3.67
	.10	3.78	3.46	3.29	3.18	3.11	3.05	2.98	2.90	2.82	2.72
	.20	2.07	2.13	2.11	2.09	2.08	2.06	2.04	2.02	1.99	1.95
7	.001	29.22	21.69	18.77	17.19	16.21	15.52	14.63	13.71	12.73	11.69
	.005	16.24	12.40	10.88	10.05	9.52	9.16	8.68	8.18	7.65	7.08
	.01	12.25	9.55	8.45	7.85	7.46	7.19	6.84	6.47	6.07	5.65
	.025	8.07	6.54	5.89	5.52	5.29	5.12	4.90	4.67	4.42	4.14
	.05	5.59	4.74	4.35	4.12	3.97	3.87	3.73	3.57	3.41	3.23
	.10	3.59	3.26	3.07	2.96	2.88	2.83	2.75	2.67	2.58	2.47
	.20	2.00	2.04	2.02	1.99	1.97	1.96	1.93	1.91	1.87	1.83
8	.001	25.42	18.49	15.83	14.39	13.49	12.86	12.04	11.19	10.30	9.34
	.005	14.69	11.04	9.60	8.81	8.30	7.95	7.50	7.01	6.50	5.95
	.01	11.26	8.65	7.59	7.01	6.63	6.37	6.03	5.67	5.28	4.86
	.025	7.57	6.06	5.42	5.05	4.82	4.65	4.43	4.20	3.95	3.67
	.05	5.32	4.46	4.07	3.84	3.69	3.58	3.44	3.28	3.12	2.93
	.10	3.46	3.11	2.92	2.81	2.73	2.67	2.59	2.50	2.40	2.29
	.20	1.95	1.98	1.95	1.92	1.90	1.88	1.86	1.83	1.79	1.74
9	.001	22.86	16.39	13.90	12.56	11.71	11.13	10.37	9.57	8.72	7.81
	.005	13.61	10.11	8.72	7.96	7.47	7.13	6.69	6.23	5.73	5.19
	.01	10.56	8.02	6.99	6.42	6.06	5.80	5.47	5.11	4.73	4.31
	.025	7.21	5.71	5.08	4.72	4.48	4.32	4.10	3.87	3.61	3.33
	.05	5.12	4.26	3.86	3.63	3.48	3.37	3.23	3.07	2.90	2.71
	.10	3.36	3.01	2.81	2.69	2.61	2.55	2.47	2.38	2.28	2.16
	.20	1.91	1.94	1.90	1.87	1.85	1.83	1.80	1.76	1.72	1.67

Table A.4. (*Continued*)

df_2 \ df_1	p	1	2	3	4	5	6	8	12	24	∞
10	.001	21.04	14.91	12.55	11.28	10.48	9.92	9.20	8.45	7.64	6.76
	.005	12.83	9.43	8.08	7.34	6.87	6.54	6.12	5.66	5.17	4.64
	.01	10.04	7.56	6.55	5.99	5.64	5.39	5.06	4.71	4.33	3.91
	.025	6.94	5.46	4.83	4.47	4.24	4.07	3.85	3.62	3.37	3.08
	.05	4.96	4.10	3.71	3.48	3.33	3.22	3.07	2.91	2.74	2.54
	.10	3.28	2.92	2.73	2.61	2.52	2.46	2.38	2.28	2.18	2.06
	.20	1.88	1.90	1.86	1.83	1.80	1.78	1.75	1.72	1.67	1.62
11	.001	19.69	13.81	11.56	10.35	9.58	9.05	8.35	7.63	6.85	6.00
	.005	12.23	8.91	7.60	6.88	6.42	6.10	5.68	5.24	4.76	4.23
	.01	9.65	7.20	6.22	5.67	5.32	5.07	4.74	4.40	4.02	3.60
	.025	6.72	5.26	4.63	4.28	4.04	3.88	3.66	3.43	3.17	2.88
	.05	4.84	3.98	3.59	3.36	3.20	3.09	2.95	2.79	2.61	2.40
	.10	3.23	2.86	2.66	2.54	2.45	2.39	2.30	2.21	2.10	1.97
	.20	1.86	1.87	1.83	1.80	1.77	1.75	1.72	1.68	1.63	1.57
12	.001	18.64	12.97	10.80	9.63	8.89	8.38	7.71	7.00	6.25	5.42
	.005	11.75	8.51	7.23	6.52	6.07	5.76	5.35	4.91	4.43	3.90
	.01	9.33	6.93	5.95	5.41	5.06	4.82	4.50	4.16	3.78	3.36
	.025	6.55	5.10	4.47	4.12	3.89	3.73	3.51	3.28	3.02	2.72
	.05	4.75	3.88	3.49	3.26	3.11	3.00	2.85	2.69	2.50	2.30
	.10	3.18	2.81	2.61	2.48	2.39	2.33	2.24	2.15	2.04	1.90
	.20	1.84	1.85	1.80	1.77	1.74	1.72	1.69	1.65	1.60	1.54
13	.001	17.81	12.31	10.21	9.07	8.35	7.86	7.21	6.52	5.78	4.97
	.005	11.37	8.19	6.93	6.23	5.79	5.48	5.08	4.64	4.17	3.65
	.01	9.07	6.70	5.74	5.20	4.86	4.62	4.30	3.96	3.59	3.16
	.025	6.41	4.97	4.35	4.00	3.77	3.60	3.39	3.15	2.89	2.60
	.05	4.67	3.80	3.41	3.18	3.02	2.92	2.77	2.60	2.42	2.21
	.10	3.14	2.76	2.56	2.43	2.35	2.28	2.20	2.10	1.98	1.85
	.20	1.82	1.83	1.78	1.75	1.72	1.69	1.66	1.62	1.57	1.51
14	.001	17.14	11.78	9.73	8.62	7.92	7.43	6.80	6.13	5.41	4.60
	.005	11.06	7.92	6.68	6.00	5.56	5.26	4.86	4.43	3.96	3.44
	.01	8.86	6.51	5.56	5.03	4.69	4.46	4.14	3.80	3.43	3.00
	.025	6.30	4.86	4.24	3.89	3.66	3.50	3.29	3.05	2.79	2.49
	.05	4.60	3.74	3.34	3.11	2.96	2.85	2.70	2.53	2.35	2.13
	.10	3.10	2.73	2.52	2.39	2.31	2.24	2.15	2.05	1.94	1.80
	.20	1.81	1.81	1.76	1.73	1.70	1.67	1.64	1.60	1.55	1.48
15	.001	16.59	11.34	9.34	8.25	7.57	7.09	6.47	5.81	5.10	4.31
	.005	10.80	7.70	6.48	5.80	5.37	5.07	4.67	4.25	3.79	3.26
	.01	8.68	6.36	5.42	4.89	4.56	4.32	4.00	3.67	3.29	2.87
	.025	6.20	4.77	4.15	3.80	3.58	3.41	3.20	2.96	2.70	2.40
	.05	4.54	3.68	3.29	3.06	2.90	2.79	2.64	2.48	2.29	2.07
	.10	3.07	2.70	2.49	2.36	2.27	2.21	2.12	2.02	1.90	1.76
	.20	1.80	1.79	1.75	1.71	1.68	1.66	1.62	1.58	1.53	1.46
16	.001	16.12	10.97	9.00	7.94	7.27	6.81	6.19	5.55	4.85	4.06
	.005	10.58	7.51	6.30	5.64	5.21	4.91	4.52	4.10	3.64	3.11
	.01	8.53	6.23	5.29	4.77	4.44	4.20	3.89	3.55	3.18	2.75
	.025	6.12	4.69	4.08	3.73	3.50	3.34	3.12	2.89	2.63	2.32
	.05	4.49	3.63	3.24	3.01	2.85	2.74	2.59	2.42	2.24	2.01
	.10	3.05	2.67	2.46	2.33	2.24	2.18	2.09	1.99	1.87	1.72
	.20	1.79	1.78	1.74	1.70	1.67	1.64	1.61	1.56	1.51	1.43
17	.001	15.72	10.66	8.73	7.68	7.02	6.56	5.96	5.32	4.63	3.85
	.005	10.38	7.35	6.16	5.50	5.07	4.78	4.39	3.97	3.51	2.98
	.01	8.40	6.11	5.18	4.67	4.34	4.10	3.79	3.45	3.08	2.65
	.025	6.04	4.62	4.01	3.66	3.44	3.28	3.06	2.82	2.56	2.25
	.05	4.45	3.59	3.20	2.96	2.81	2.70	2.55	2.38	2.19	1.96
	.10	3.03	2.64	2.44	2.31	2.22	2.15	2.06	1.96	1.84	1.69
	.20	1.78	1.77	1.72	1.68	1.65	1.63	1.59	1.55	1.49	1.42
18	.001	15.38	10.39	8.49	7.46	6.81	6.35	5.76	5.13	4.45	3.67
	.005	10.22	7.21	6.03	5.37	4.96	4.66	4.28	3.86	3.40	2.87
	.01	8.28	6.01	5.09	4.58	4.25	4.01	3.71	3.37	3.00	2.57
	.025	5.98	4.56	3.95	3.61	3.38	3.22	3.01	2.77	2.50	2.19
	.05	4.41	3.55	3.16	2.93	2.77	2.66	2.51	2.34	2.15	1.92
	.10	3.01	2.62	2.42	2.29	2.20	2.13	2.04	1.93	1.81	1.66
	.20	1.77	1.76	1.71	1.67	1.64	1.62	1.58	1.53	1.48	1.40

Table A.4. (*Continued*)

df_2 \ df_1	p	1	2	3	4	5	6	8	12	24	∞
19	.001	15.08	10.16	8.28	7.26	6.61	6.18	5.59	4.97	4.29	3.52
	.005	10.07	7.09	5.92	5.27	4.85	4.56	4.18	3.76	3.31	2.78
	.01	8.18	5.93	5.01	4.50	4.17	3.94	3.63	3.30	2.92	2.49
	.025	5.92	4.51	3.90	3.56	3.33	3.17	2.96	2.72	2.45	2.13
	.05	4.38	3.52	3.13	2.90	2.74	2.63	2.48	2.31	2.11	1.88
	.10	2.99	2.61	2.40	2.27	2.18	2.11	2.02	1.91	1.79	1.63
	.20	1.76	1.75	1.70	1.66	1.63	1.61	1.57	1.52	1.46	1.39
20	.001	14.82	9.95	8.10	7.10	6.46	6.02	5.44	4.82	4.15	3.38
	.005	9.94	6.99	5.82	5.17	4.76	4.47	4.09	3.68	3.22	2.69
	.01	8.10	5.85	4.94	4.43	4.10	3.87	3.56	3.23	2.86	2.42
	.025	5.87	4.46	3.86	3.51	3.29	3.13	2.91	2.68	2.41	2.09
	.05	4.35	3.49	3.10	2.87	2.71	2.60	2.45	2.28	2.08	1.84
	.10	2.97	2.59	2.38	2.25	2.16	2.09	2.00	1.89	1.77	1.61
	.20	1.76	1.75	1.70	1.65	1.62	1.60	1.56	1.51	1.45	1.37
21	.001	14.59	9.77	7.94	6.95	6.32	5.88	5.31	4.70	4.03	3.26
	.005	9.83	6.89	5.73	5.09	4.68	4.39	4.01	3.60	3.15	2.61
	.01	8.02	5.78	4.87	4.37	4.04	3.81	3.51	3.17	2.80	2.36
	.025	5.83	4.42	3.82	3.48	3.25	3.09	2.87	2.64	2.37	2.04
	.05	4.32	3.47	3.07	2.84	2.68	2.57	2.42	2.25	2.05	1.81
	.10	2.96	2.57	2.36	2.23	2.14	2.08	1.98	1.88	1.75	1.59
	.20	1.75	1.74	1.69	1.65	1.61	1.59	1.55	1.50	1.44	1.36
22	.001	14.38	9.61	7.80	6.81	6.19	5.76	5.19	4.58	3.92	3.15
	.005	9.73	6.81	5.65	5.02	4.61	4.32	3.94	3.54	3.08	2.55
	.01	7.94	5.72	4.82	4.31	3.99	3.76	3.45	3.12	2.75	2.31
	.025	5.79	4.38	3.78	3.44	3.22	3.05	2.84	2.60	2.33	2.00
	.05	4.30	3.44	3.05	2.82	2.66	2.55	2.40	2.23	2.03	1.78
	.10	2.95	2.56	2.35	2.22	2.13	2.06	1.97	1.86	1.73	1.57
	.20	1.75	1.73	1.68	1.64	1.61	1.58	1.54	1.49	1.43	1.35
23	.001	14.19	9.47	7.67	6.69	6.08	5.65	5.09	4.48	3.82	3.05
	.005	9.63	6.73	5.58	4.95	4.54	4.26	3.88	3.47	3.02	2.48
	.01	7.88	5.66	4.76	4.26	3.94	3.71	3.41	3.07	2.70	2.26
	.025	5.75	4.35	3.75	3.41	3.18	3.02	2.81	2.57	2.30	1.97
	.05	4.28	3.42	3.03	2.80	2.64	2.53	2.38	2.20	2.00	1.76
	.10	2.94	2.55	2.34	2.21	2.11	2.05	1.95	1.84	1.72	1.55
	.20	1.74	1.73	1.68	1.63	1.60	1.57	1.53	1.49	1.42	1.34
24	.001	14.03	9.34	7.55	6.59	5.98	5.55	4.99	4.39	3.74	2.97
	.005	9.55	6.66	5.52	4.89	4.49	4.20	3.83	3.42	2.97	2.43
	.01	7.82	5.61	4.72	4.22	3.90	3.67	3.36	3.03	2.66	2.21
	.025	5.72	4.32	3.72	3.38	3.15	2.99	2.78	2.54	2.27	1.94
	.05	4.26	3.40	3.01	2.78	2.62	2.51	2.36	2.18	1.98	1.73
	.10	2.93	2.54	2.33	2.19	2.10	2.04	1.94	1.83	1.70	1.53
	.20	1.74	1.72	1.67	1.63	1.59	1.57	1.53	1.48	1.42	1.33
25	.001	13.88	9.22	7.45	6.49	5.88	5.46	4.91	4.31	3.66	2.89
	.005	9.48	6.60	5.46	4.84	4.43	4.15	3.78	3.37	2.92	2.38
	.01	7.77	5.57	4.68	4.18	3.86	3.63	3.32	2.99	2.62	2.17
	.025	5.69	4.29	3.69	3.35	3.13	2.97	2.75	2.51	2.24	1.91
	.05	4.24	3.38	2.99	2.76	2.60	2.49	2.34	2.16	1.96	1.71
	.10	2.92	2.53	2.32	2.18	2.09	2.02	1.93	1.82	1.69	1.52
	.20	1.73	1.72	1.66	1.62	1.59	1.56	1.52	1.47	1.41	1.32
26	.001	13.74	9.12	7.36	6.41	5.80	5.38	4.83	4.24	3.59	2.82
	.005	9.41	6.54	5.41	4.79	4.38	4.10	3.73	3.33	2.87	2.33
	.01	7.72	5.53	4.64	4.14	3.82	3.59	3.29	2.96	2.58	2.13
	.025	5.66	4.27	3.67	3.33	3.10	2.94	2.73	2.49	2.22	1.88
	.05	4.22	3.37	2.98	2.74	2.59	2.47	2.32	2.15	1.95	1.69
	.10	2.91	2.52	2.31	2.17	2.08	2.01	1.92	1.81	1.68	1.50
	.20	1.73	1.71	1.66	1.62	1.58	1.56	1.52	1.47	1.40	1.31
27	.001	13.61	9.02	7.27	6.33	5.73	5.31	4.76	4.17	3.52	2.75
	.005	9.34	6.49	5.36	4.74	4.34	4.06	3.69	3.28	2.83	2.29
	.01	7.68	5.49	4.60	4.11	3.78	3.56	3.26	2.93	2.55	2.10
	.025	5.63	4.24	3.65	3.31	3.08	2.92	2.71	2.47	2.19	1.85
	.05	4.21	3.35	2.96	2.73	2.57	2.46	2.30	2.13	1.93	1.67
	.10	2.90	2.51	2.30	2.17	2.07	2.00	1.91	1.80	1.67	1.49
	.20	1.73	1.71	1.66	1.61	1.58	1.55	1.51	1.46	1.40	1.30

Table A.4. (*Continued*)

df_2 \ df_1	p	1	2	3	4	5	6	8	12	24	∞
28	.001	13.50	8.93	7.19	6.25	5.66	5.24	4.69	4.11	3.46	2.70
	.005	9.28	6.44	5.32	4.70	4.30	4.02	3.65	3.25	2.79	2.25
	.01	7.64	5.45	4.57	4.07	3.75	3.53	3.23	2.90	2.52	2.06
	.025	5.61	4.22	3.63	3.29	3.06	2.90	2.69	2.45	2.17	1.83
	.05	4.20	3.34	2.95	2.71	2.56	2.44	2.29	2.12	1.91	1.65
	.10	2.89	2.50	2.29	2.16	2.06	2.00	1.90	1.79	1.66	1.48
	.20	1.72	1.71	1.65	1.61	1.57	1.55	1.51	1.46	1.39	1.30
29	.001	13.39	8.85	7.12	6.19	5.59	5.18	4.64	4.05	3.41	2.64
	.005	9.23	6.40	5.28	4.66	4.26	3.98	3.61	3.21	2.76	2.21
	.01	7.60	5.42	4.54	4.04	3.73	3.50	3.20	2.87	2.49	2.03
	.025	5.59	4.20	3.61	3.27	3.04	2.88	2.67	2.43	2.15	1.81
	.05	4.18	3.33	2.93	2.70	2.54	2.43	2.28	2.10	1.90	1.64
	.10	2.89	2.50	2.28	2.15	2.06	1.99	1.89	1.78	1.65	1.47
	.20	1.72	1.70	1.65	1.60	1.57	1.54	1.50	1.45	1.39	1.29
30	.001	13.29	8.77	7.05	6.12	5.53	5.12	4.58	4.00	3.36	2.59
	.005	9.18	6.35	5.24	4.62	4.23	3.95	3.58	3.18	2.73	2.18
	.01	7.56	5.39	4.51	4.02	3.70	3.47	3.17	2.84	2.47	2.01
	.025	5.57	4.18	3.59	3.25	3.03	2.87	2.65	2.41	2.14	1.79
	.05	4.17	3.32	2.92	2.69	2.53	2.42	2.27	2.09	1.89	1.62
	.10	2.88	2.49	2.28	2.14	2.05	1.98	1.88	1.77	1.64	1.46
	.20	1.72	1.70	1.64	1.60	1.57	1.54	1.50	1.45	1.38	1.28
40	.001	12.61	8.25	6.60	5.70	5.13	4.73	4.21	3.64	3.01	2.23
	.005	8.83	6.07	4.98	4.37	3.99	3.71	3.35	2.95	2.50	1.93
	.01	7.31	5.18	4.31	3.83	3.51	3.29	2.99	2.66	2.29	1.80
	.025	5.42	4.05	3.46	3.13	2.90	2.74	2.53	2.29	2.01	1.64
	.05	4.08	3.23	2.84	2.61	2.45	2.34	2.18	2.00	1.79	1.51
	.10	2.84	2.44	2.23	2.09	2.00	1.93	1.83	1.71	1.57	1.38
	.20	1.70	1.68	1.62	1.57	1.54	1.51	1.47	1.41	1.34	1.24
60	.001	11.97	7.76	6.17	5.31	4.76	4.37	3.87	3.31	2.69	1.90
	.005	8.49	5.80	4.73	4.14	3.76	3.49	3.13	2.74	2.29	1.69
	.01	7.08	4.98	4.13	3.65	3.34	3.12	2.82	2.50	2.12	1.60
	.025	5.29	3.93	3.34	3.01	2.79	2.63	2.41	2.17	1.88	1.48
	.05	4.00	3.15	2.76	2.52	2.37	2.25	2.10	1.92	1.70	1.39
	.10	2.79	2.39	2.18	2.04	1.95	1.87	1.77	1.66	1.51	1.29
	.20	1.68	1.65	1.59	1.55	1.51	1.48	1.44	1.38	1.31	1.18
120	.001	11.38	7.31	5.79	4.95	4.42	4.04	3.55	3.02	2.40	1.56
	.005	8.18	5.54	4.50	3.92	3.55	3.28	2.93	2.54	2.09	1.43
	.01	6.85	4.79	3.95	3.48	3.17	2.96	2.66	2.34	1.95	1.38
	.025	5.15	3.80	3.23	2.89	2.67	2.52	2.30	2.05	1.76	1.31
	.05	3.92	3.07	2.68	2.45	2.29	2.17	2.02	1.83	1.61	1.25
	.10	2.75	2.35	2.13	1.99	1.90	1.82	1.72	1.60	1.45	1.19
	.20	1.66	1.63	1.57	1.52	1.48	1.45	1.41	1.35	1.27	1.12
∞	.001	10.83	6.91	5.42	4.62	4.10	3.74	3.27	2.74	2.13	1.00
	.005	7.88	5.30	4.28	3.72	3.35	3.09	2.74	2.36	1.90	1.00
	.01	6.64	4.60	3.78	3.32	3.02	2.80	2.51	2.18	1.79	1.00
	.025	5.02	3.69	3.12	2.79	2.57	2.41	2.19	1.94	1.64	1.00
	.05	3.84	2.99	2.60	2.37	2.21	2.09	1.94	1.75	1.52	1.00
	.10	2.71	2.30	2.08	1.94	1.85	1.77	1.67	1.55	1.38	1.00
	.20	1.64	1.61	1.55	1.50	1.46	1.43	1.38	1.32	1.23	1.00

Source: Reproduced from E. F. Lindquist, *Design and Analysis of Experiments in Psychology and Education,* Houghton, Mifflin, Boston, 1953, pp. 41–44, with permission of the publisher.

References

Abelson, R. P. *Testing* a priori *hypotheses in the analysis of variance.* Unpublished manuscript, Yale University, New Haven, 1962.

American Psychological Association. *Publication Manual of the American Psychological Association* (3rd ed.). Washington, DC: Author, 1983.

Armor, D. J. & Couch, A. S. *Data-text primer: An introduction to computerized social data analysis.* NY: Free Press, 1972.

Balaam, L. N. Multiple comparisons – A sampling experiment. *Australian Journal of Statistics*, 1963, *5*, 62–84.

Battig, W. F. Parsimony in psychology. *Psychological Reports*, 1962, *11*, 555–572.

Box, G. E. P. Non-normality and tests on variances. *Biometrika*, 1953, *40*, 318–335.

Cochran, W. G. The comparison of percentages in matched samples. *Biometrika*, 1950, *37*, 256–266.

Cohen, J. *Statistical power analysis for the behavioral sciences.* NY: Academic Press, 1969.

Statistical power analysis for the behavioral sciences (rev. ed.). NY: Academic Press, 1977.

Converse, J. Issue-importance and forced-compliance attitude change: Another curvilinear finding. *Personality and Social Psychology Bulletin*, 1982, *8*, 651–655.

Converse, J. & Cooper, J. The importance of decisions and free-choice attitude change: A curvilinear finding. *Journal of Experimental Social Psychology*, 1979, *15*, 48–61.

D'Agostino, R. B. A second look at analysis of variance on dichotomous data. *Journal of Educational Measurement*, 1971, *8*, 327–333.

Eagly, A. H. Recipient characteristics as determinants of responses to persuasion. In R. E. Petty, T. M. Ostrom, & T. C. Brock (eds.), *Cognitive responses in persuasion.* Hillsdale, NJ: Erlbaum, 1981, pp. 173–195.

Edwards, A. L. *Experimental design in psychological research* (4th ed.). NY: Holt, Rinehart and Winston, 1972.

Grant, D. A. Analysis-of-variance tests in the analysis and comparison of curves. *Psychological Bulletin*, 1956, *53*, 141–154.

Green, B. F., Jr. & Tukey, J. W. Complex analysis of variance: General problems. *Psychometrika*, 1960, *7*, 127–152.

Harris, R. J. *A primer of multivariate statistics.* New York: Academic Press, 1975.

Hays, W. L. *Statistics* (3rd ed.). NY: Holt, Rinehart and Winston, 1981.

Levin, S. Behind every great man is a woman, behind every great woman there is none: A look at *Who's Who in America.* Unpublished data, Cambridge, MA: Harvard University, 1974.

Luchins, A. S. & Luchins, E. H. *Logical foundations of mathematics for behavioral scientists.* NY: Holt, Rinehart and Winston, 1965.

Lunney, G. H. Using analysis of variance with a dichotomous dependent variable: An empirical study. *Journal of Educational Measurement*, 1970, *7*, 263–269.

McGuire, W. J. Personality and susceptibility to social influence. In E. F. Borgatta & W. W. Lambert (eds.), *Handbook of personality theory and research.* Chicago: Rand-McNally, 1968.

Malmo, R. B. Activation: A neuropsychological dimension. *Psychological Review*, 1959, *66*, 367–386.

Morrison, D. F. *Multivariate statistical methods* (2nd ed.). NY: McGraw-Hill, 1976.

Myers, J. L. *Fundamentals of experimental design* (3rd ed.). Boston: Allyn and Bacon, 1979.

Rosenthal, R. *Experimenter effects in behavioral research.* NY: Appleton-Century-Crofts, 1966.

Interpersonal expectations: Effects of the experimenter's hypothesis. In R. Rosenthal & R. L. Rosnow (eds.), *Artifact in behavioral research.* NY: Academic Press, 1969.

Experimenter effects in behavioral research (enl. ed.). NY: Halsted Press (Irvington), 1976.

Rosenthal, R. (Ed.). *New directions for methodology of social and behavioral science: Quantitative assessment of research domains.* San Francisco: Jossey-Bass, 1980.

Pavlov's mice, Pfungst's horse, and Pygmalion's PONS: Some models for the study of interpersonal expectancy effects. In T. A. Sebeok & R. Rosenthal (eds.), *The Clever Hans phenomenon: Communication with horses, whales, apes, and people. Annals of the New York Academy of Sciences*, 1981, No. 364, pp. 182–198.

(In press). From unconscious experimenter bias to teacher expectancy effects. In J. B. Dusek, V. C. Hall, & W. J. Meyer (eds.), *Teacher expectancies.* Hillsdale, NJ: Lawrence Erlbaum Associates.

Meta-analytic procedures in social research. Beverly Hills, CA: Sage, 1984.

Rosenthal, R., Hall, J. A., DiMatteo, M. R., Rogers, P. L., & Archer, D. *Sensitivity to nonverbal communication: The PONS test.* Baltimore, MD: The Johns Hopkins University Press, 1979.

Rosenthal, R. & Jacobson, L. *Pygmalion in the classroom: Teacher expectation and pupil's intellectual development.* NY: Holt, Rinehart and Winston, 1968.

Rosenthal, R. & Rosnow, R. L. *Essentials of behavioral research: Methods and data analysis.* NY: McGraw-Hill, 1984.

Rosenthal, R. & Rubin, D. B. Interpersonal expectancy effects: The first 345 studies. *The Behavioral and Brain Sciences,* 1978a, *3,* 377–386.

Issues in summarizing the first 345 studies of interpersonal expectancy effects. *The Behavioral and Brain Sciences,* 1978b, *3,* 410–415.

Comparing within- and between-subjects studies. *Sociological Methods and Research,* 1980, *9,* 127–136.

Ensemble-adjusted *p* values. *Psychological Bulletin,* 1983, *94,* 540–541.

Multiple contrasts and ordered Bonferroni procedures. *Journal of Educational Psychology,* 1984, *76,* 1028–1034.

Rosnow, R. L. The prophetic vision of Giambattista Vico: Implications for the state of social psychological theory. *Journal of Personality and Social Psychology,* 1978, *36,* 1322–1331.

Psychology of rumor reconsidered. *Psychological Bulletin,* 1980, *87,* 578–591.

Paradigms in transition: The methodology of social inquiry. NY: Oxford University Press, 1981.

Von Osten's horse, Hamlet's question, and the mechanistic view of causality: Implications for a post-crisis social psychology. *Journal of Mind and Behavior,* 1983, *4,* 319–338.

Schultz, D. P. *Sensory restriction: Effects on behavior.* NY: Academic Press, 1965.

Snedecor, G. W. & Cochran, W. G. *Statistical methods* (6th ed.). Ames, Iowa: Iowa State University Press, 1967.

Statistical methods (7th ed.). Ames, Iowa: Iowa State University Press, 1980.

Thorndike, E. L. Fundamental theorems in judging men. *Journal of Applied Psychology,* 1918, *2,* 67–76.

Webb, E. J., Campbell, D. T., Schwartz, R. D., and Sechrest, L. *Unobtrusive measures: Nonreactive research in the social sciences.* Chicago: Rand McNally, 1966.

Winer, B. J. *Statistical principles in experimental design* (2nd ed.). 1971, NY: McGraw-Hill.

Index